Academic Environment

Academic Environment

A Handbook for Evaluating Employment Opportunities in Science

Second Edition

Karl W. Lanks

Director, Department of Pathology and Laboratory Medicine
Staten Island University Hospital
Staten Island, New York

CRC Press
Taylor & Francis Group
Boca Raton London New York

CRC Press is an imprint of the
Taylor & Francis Group, an **informa** business

Contents

Part Three
Objective Evaluation of the Academic Environment

Preface

Intimidated by the specter of the taxi-driving Ph.D., university faculty have considered themselves lucky to have any sort of salaried position and have rarely had the luxury of choice or the leverage to negotiate. As the first edition of the handbook took shape, that picture seemed to be changing rapidly. An analysis of university faculty requirements conducted by W. G. Bowen and J. A. Sousa indicated that within the next 10 years the number of job openings in all arts and sciences would exceed the supply of faculty candidates.[1] This dramatic reversal of supply and demand would place scholars in a position they had not enjoyed since the 1960s—that of being able to choose among competing institutions and negotiate the terms of their employment.

Shortly after publication, a series of unforeseen events wracked the academic job market. The most severe economic down turn since the Great Depression forced universities to curtail hiring. Pharmaceutical companies abruptly changed their business strategy to deemphasize in-house basic research. Finally (and this process continues), reduced staff requirements and revenue expectations under "managed care" are leading to wholesale migration of Faculty of the American Society of Experimental Biologists members out of many traditional academic centers. Nevertheless, well-trained faculty and staff are still in demand, large numbers of positions are still being filled each year, and the overall employment prospects are no worse now than in the early 1980s.

Identifying an institutional environment conducive to career development and personal satisfaction can be one of the most difficult problems faced by any scholar. Many questions are asked, especially when considering an employment offer. "Can I live like a professor on this salary?" "Should I expect this contract to be honored?" "How much difference does locating at a particular institution make anyway?" Unable to refer to an objective database, each individual arrives at a personal answer after informally evaluating institutional atmosphere and working conditions. The conclusions reflect traditional assumptions, hearsay, and biased opinions that may be far from the truth.

The problem is further compounded by the value system that becomes an integral part of every academician's thinking. From classicist to scientist, all are trained to analyze data honestly, to see many sides of a problem, and to trust the validity of work that has been done and verified by others whom they probably do not know personally. This background does not prepare them for situations in which deceit, egomania, racial or sexual bias, and political expediency take precedence over objectivity.

This is not to say that academicians are naive or lack experience in the "real" world. No one would consider lawyers naive because they lack basic concepts about Elizabethan pronunciation or expression vectors. Similarly, professors are not naive simply because they know little or nothing about contracts. Each is simply specialized in a different area of human endeavor. Such specialization is not a problem for lawyers, but it is for scientists. This is so because, although lawyers generally do not recite Shakespeare or clone genes, professors do write contracts.

This handbook deals with contracts, evaluating job offers, lawsuits, and deciding when to leave an institution. The relationship between institutional affiliation and productivity is also discussed in detail. These are matters of concern to all college and university faculty from part-time adjunct to tenured full professors as well as to industrial scientists. Other aspects of career development include building a curriculum vitae and developing a network of colleagues. These professional activities are essential prerequisites for any aspiring scientist, but are so strongly influenced by the unique culture and expectations of individual disciplines that they cannot be considered in detail. Therefore, the handbook begins with the job offer and has the following objectives.

Objective 1 To help scholars avoid problems that interfere with career development and to deal with such problems if they do arise.

Objective 2 To provide an objective system that can be used to evaluate the atmosphere and working conditions in any academic or industrial environment.

The evaluation system described in Chapter 6 is based on responses to a random survey of full-time faculty or staff at medical schools, major universities, selected liberal arts colleges, and pharmaceutical companies in the

United States and Canada. Even though biomedically oriented institutions were surveyed, most of the questions asked (see Appendix I) were clearly applicable to any institution of higher learning or industrial research organization. Readers interested in evaluating institutions that were not surveyed will have no difficulty using these questions and Chapter 2 as a guide.

By and large, the 1562 respondents to the survey represent senior faculty or professional staff who have been at their current institution for more than 10 years.[2] These faculty should know what they are talking about and had no reason to lie: they had sufficient opportunity to encounter most of the usual problems, were tenured (or the equivalent), and were responding anonymously. Nevertheless, individual departments vary within an institution and change with time, so the survey results and the rankings derived from them should serve as a starting point for a more detailed investigation of specific niches rather than as absolute and sufficient criteria in themselves.

Given these caveats, 95% of the respondents felt that the choice of institution had affected their career development either positively or negatively in one of the areas covered by the survey. Such strong opinions indicated that faculty and professional staff consider institutional affiliation to be a major determinant of career satisfaction. A comprehensive evaluation of academic atmosphere taking into account career development, faculty-administration relations, and other subjective factors showed that faculty and professional staff opinions were remarkably consistent for any one institution, but differed significantly among institutions. Thus, about one-third of institutions surveyed were judged to have a highly positive general academic atmosphere. Another one-fifth were judged to have a highly negative atmosphere and the remainder formed a group in which the academic atmosphere evaluation was not strongly polarized one way or the other. These results can now be used to identify institutions with high levels of faculty career satisfaction.

Subjective assessment of job satisfaction is very important, but faculty and professional staff also want to know whether locating at a specific institution will affect their professional productivity. This problem is analyzed by asking whether institutional and per capita production of grants and publications differed within the group of institutions surveyed for academic atmosphere. This analysis leads to the conclusion that productivity of individual faculty is relatively constant from school to school with differences in faculty size accounting for the apparent differences in productivity among the institutions. The aspects of academic atmosphere that characterize large, highly productive schools are identified, and the analysis is extended to explain how institutional administration affects productivity.

The approach taken in this handbook applies objective analysis to solve, or at least clarify, problems that are usually considered psychosocial. That the results are novel should not be too surprising because all the underlying data were compiled specifically for this project and are not available in any other form. However, the problems themselves are not new. Quotations from Confucius (551–479 B.C.) at the beginning of each chapter show that scholars have been dealing with similar questions for over 2000 years.[3] This handbook aims to help academicians and industrial scientists make good use of the progress that has been made.

Karl W. Lanks, M.D., Ph.D.
Staten Island, New York

PART I

THE ACADEMIC
ENVIRONMENT

Chapter 1

The Offer

If you had a piece of beautiful jade here, would you put it away safely in a box or would you try to sell it for a good price? The Master said, "Of course I would sell it. Of course I would sell it. All I am waiting for is the right offer." Analects, IX.13.

The relief and excitement afforded by a firm employment offer after at least 15 years of preparation accounts for the reluctance of many young scholars and scientists to examine the details critically. They are simply afraid that probing too deeply will ruin the deal and leave them without a job. These concerns will be dealt with later, but some excessively cynical overkill is needed at this point to set the tone for the more objective discussion to follow.

Naturally, you want to assume that you were offered the position because you were the most meritorious candidate. Nothing could be farther from the truth. At best, you are the cheapest of several alternatives. Why is recruiting not being done at a higher level, say full professor? Of course cost is a factor; any competent department chairperson or director will seek to satisfy institutional requirements with the smallest possible outlay. At worst, you are seen as a potential sucker who might fall for a deal that no one else will accept. Intermediate possibilities are also distasteful, but you get the picture.

How then to reconcile the euphoria of what should be a genuinely happy event with the daunting realization of potential treachery? The answer is simply to follow the three rules for evaluating an offer.

EVALUATING AN OFFER

Rule 1 Investigate the chairperson/director.
Rule 2 Investigate the department.
Rule 3 Investigate the institution.

EVALUATING THE CHAIRPERSON/DIRECTOR

No job offer (or any other statement for that matter) can be fully understood and properly weighed unless the motivation behind it is also known. Chairpersons/directors come in innumerable styles, but only a few sources of motivation are strong enough to induce anyone to take such a thankless job. These are altruism, greed, ego, and survival instinct: Corresponding chairperson types are Altruist, Parasite, Egoist, and Survivor. A leader motivated by one or another of these drives can affect departmental atmosphere and staff career development so strongly that this individual's personality must be considered part of the offer package. Most of the following discussion is devoted to identifying the latter three types, not because they are necessarily the most common, but because their potential for damage is so great.

Altruists exist and some devote their lives to building strong academic departments and industrial research organizations in which faculty or professional staff who measure up to their standards can flourish. The general atmosphere can be highly positive with superb opportunities for career development. On the other hand, individuals with psychosocial problems that interfere with personal relations or intellectual inadequacies that limit their productivity will not last long in such departments. Given the wide range of standards (only 5% of departments can be in the top 5%) and the alternatives described below, this sort of department approaches the ideal.

In the process of identifying the Altruist it is essential that the other three types be eliminated. Although not sufficient for identification, it also helps to note that the department will probably be very good, given the available resources, with clearly stated objectives for future development that are consistent with general trends in the field. The track record (see below) will indicate sustained and substantial contribution in a specific area of expertise and proven ability to develop the careers of subordinates.

Chairs, especially clinical chairs, carry a higher salary than professorships and no one expects the occupants to donate the difference to charity. However, leaders motivated predominantly by greed use their position primarily for personal aggrandizement, hence the designation Parasite. Parasitism is also manifest when most of the departmental resources are appropriated to support the chairperson's laboratory even though personal wealth is not involved. You can flourish in this environment by being a "good" worker, i.e., by bringing in more practice income or by making the lab more productive. Don't expect to get most of the credit, but if your share is negotiable this may not be too bad an arrangement. After you pay your dues and become independent, you can always leave (see Chapter 5).

What better way to shore up the Egoist's tottering sense of personal worth than by taking an academic chair. Title is of primary importance; prestige and reputation of the department are secondary. If the individuals are intelligent and competent, only insecure, they may be indistinguishable from altruists and their thoughts are their own business. However, if insecurity has driven them beyond their abilities, then challenges to intellect and administrative skill can be met only with arrogance and aggression. Concerns that should be kept between them and their psychiatrists become everyone else's as well.

Egoists cannot allow subordinates to flourish because the success of others is implicitly threatening. However, you can exist in this sort of environment and even get promoted by submerging your own ego, minimizing any outward appearance of success, avoiding all conflict, and using flattery so blatant that it would be disgusting to anyone else.

Observation is the key to recognizing the Egoist. Are people in the lab treated like servants or put down in your presence? Do staff display initiative in the presence of this person? Are entrances, particularly at first meeting, intentionally dramatic? As pointed out below, there are also specific individuals in the department who will be happy to discuss this individual's failings with you.

Finally, there are the entirely incompetent individuals who realize that they can best survive by becoming the boss. Lacking a high level of intelligence or technical skill the Survivor can achieve this goal only through deceit, treachery, and Machiavellian plotting. Survivors are typically quite articulate, using this skill to magnify their trivial contributions.

In order to prosper with this type you have two choices: lick 'em or join 'em. Considering that they are not highly intelligent, it is possible to rely on the overwhelming weight of *your* accomplishments to provide a continuing level of intimidation and to simply outsmart them as specific situations arise. However, this diverts physical and emotional energy that could be put to more productive use and creates enough tension to increase the risk of ulcer or heart disease. Curiously, other Survivor types fare the best in this environment. Perhaps their lack of scruples makes them indispensable as accomplices. It may work for now, but remember what happened to Dr. Faustus.

The Survivor can be recognized by getting a print-out of all publications during the 20 years before becoming a chairperson and crossing out the case reports, abstracts, courtesy citations, reviews, and other miscellany. Does the remainder show a period of ten consecutive years with one or two major papers per year in highly respected journals? If not, then how did this person come to occupy the chair? Most likely by virtue of the "survival skills" mentioned above.

Besides relying on your own instant recognition skills there is a cardinal principle to which anyone investigating a chairperson must adhere. You *must* speak to as many potential informants as possible. Absolutely never limit your source of information to the handpicked cronies to whom you are introduced at the initial interviews.

YOUR POTENTIAL INFORMATION SOURCES

Source 1 All departmental faculty and professional staff
Source 2 Former fellows and students
Source 3 Colleagues formerly at the institution

Ask open-ended questions such as, "What do you think working in this department will be like?" and don't expect any directly negative answers. Look for what is left unsaid. Speak to every member of the department, especially those who were on board before the present chairperson/director. Former students probably know the boss better than anyone else and are usually overlooked as a source of information. If former colleagues have little good to say, then you can bet that you won't either after a few years.

Having completed this stage of the evaluation you are now in a position to decide whether this offer might be acceptable to you. Perhaps no offer made by this particular chairperson would be acceptable and negotiations can be aborted at an early stage. If the offer still seems to hold potential, then widening the scope of your investigation is in order.

EVALUATING THE DEPARTMENT

The makeup of the rest of the department makes less difference than generally supposed. In attempting to achieve breadth in a field, most academic departments end up with faculty specialized in such widely divergent areas that they are more likely to find collaborators, and therefore their day-to-day contacts, in a different department or institution than in their own. Even relations among individuals and groups of faculty depend largely on the personality at the top. A Survivor type will set individuals and groups against each other to discourage unity and cooperation whereas other less destructive types will use a variety of tactics to avoid internal conflict.

The really important questions have to do with the adequacy of shared resources and the mechanisms by which they are allocated. For example, if four strong faculty are already competing for two graduate students, your chance of getting one is probably nil. In general, the more closely the system approaches a participatory democracy, i.e., the greater the faculty input into departmental administration, the better the work atmosphere. This probably reflects both the ability of better chairpersons to delegate authority and the salutary effects of determining one's own fate.

Professional staff of industrial research organizations face many of the same problems, but may lack the autonomy and traditional protections of their academic counterparts. They must be especially careful that their ability to rise within the organization based on their own merit is not impeded by the prospective department director. The question to ask is whether promising projects are transferred to a favorite group leader or allowed to remain with the group that initiated them. If you don't get the credit for initiating the development of potential products, you are unlikely to go far in a pharmaceutical company.

EVALUATING THE INSTITUTION

Even though new staff are not likely to interact frequently with the president or with chairpersons and directors of other departments, several aspects of the larger organization need to be considered. Potential collaborators should be identified and contacted. Faculty at institutions having highly positive effects on career development rank the availability of collaborators among the most important factors. Of course, this process also provides an excuse for talking to a few staff from outside the department about the conditions on the inside.

If you are a woman, a homosexual, or a member of a racial, religious, or ethnic minority, checking into the department's history of dealing with your group is essential. Of course, you will have spoken to anyone who is currently there, but they may be reluctant to tell the whole truth for fear of driving you away and losing a fellow sufferer. Former department members are more useful. Institutional cultures have a long half-life so anyone with your characteristics who has left within the last 20 years should be contacted. Go to the library and look up some old school catalogs: you might be surprised who has been through the place. This might be a good opportunity to get in touch with them and you can be sure that they will be anxious to relate any relevant horror stories.

The final reason for considering the supradepartmental administration has to do with the process by which administrators are "cloned." Deans and presidents choose chairpersons and directors with whom they are comfortable and who fit into the organizational pattern. Unless they make a mistake, the attitudes of all their appointees will be similar and the institution will acquire a characteristic atmosphere. Therefore, unless the institution has recently undergone a radical reorganization, you can reason from the general to the specific and deduce the atmosphere of your department from that of the institution as a whole. An objective method for evaluating institutional atmosphere is presented in Chapter 2. If this approach is not valid for your specific department, then it is up to someone to explain with some precision how this department has come to isolate itself from the rest of the institution.

Chapter 2

The Contract

He who gives no thought to difficulties in the future is sure to be beset by worries much closer at hand. Analects, XV.12.

Because the context of the offer is more important than the content, very little time should be wasted once the multilevel investigation of the preceding chapter is satisfactorily completed. There are only two questions: "Am I being promised enough?" and "How will the goods be delivered?" The needs of individual research disciplines dictate the resources required to set up an independent laboratory. Since autonomy is the goal of most bio-research faculty, no institution could hope to attract high quality faculty by offering resources inadequate to achieve this goal. Salary arrangements and delivery of the promised resources are entirely different matters.

The salaries of medical school faculty surveyed in 1989 and the effects of salary on life-style provide a useful perspective. Salaries of faculty without clinical income were only slightly higher than the $52,000 found for full professors of chemistry in degree-granting institutions.[1] When survey faculty were asked whether their family income from salary would permit a life-style appropriate to their rank and comparable to that of their peers, 62% responded "definitely" or "most likely," 21% responded "borderline," and 17% responded "not likely" or "definitely not." The negative responses were clustered in the lower income ranges, comprising 29% of the less than $50,000 group without clinical income, but only 13% of the corresponding $51,000–$75,000 group.

Thus, whereas most biomedical scientists in the study population felt that income from salary meets their expectations and requirements, a significant fraction of otherwise successful individuals felt that it did not. Because about one-half of faculty in nonmedical institutions earns less than $50,000, generalizing this result suggests that at least one-third of faculty at U.S. universities are dissatisfied with their monetary compensation.

This conclusion tends to dispel the belief that academics are content with low salaries in exchange for academic freedom and makes salary a legitimate issue in preemployment negotiations. If the survey results are used as a guide, then a salary of less than $50,000 invites future dissatisfaction and is to be avoided. Of course, various standard excuses are dragged out to justify a low salary. The most spectacularly specious is that the cost of living is so much lower in your area. In fact, differences in cost of living justify an adjustment of only about 10% in daily living expenses and even this minor adjustment fails to take into consideration problems associated with faculty mobility.

After ten years on a low salary in a "low cost of living area" you will not have saved enough to move to an area where housing prices are higher. Once residents in an area where housing prices are high purchase a home, they have hitched a ride on the inflation spiral and can afford this quality home no matter how high prices rise. The same process locks everyone else out unless they also accumulate assets in parallel. This effect is seen nowhere better than in New York City where housing costs now make it practically impossible to recruit new faculty from outside the city. Whereas you may not want to move to New York right now, most of the East and West coasts are afflicted with housing costs nearly as high.

Not only do you need the salary, but you need a means of assuring that other promised items are indeed delivered. The contract is simply a routine means by which parties formalize such promises. Three rules are to be observed.

RULES OF CONTRACTS

Rule 1 Get a contract.
Rule 2 Get a written contract.
Rule 3 Get a detailed written contract.

If a man's word is his bond (as the saying goes), then why do I need a contract? Questions like this one, which embrace four distinct misconceptions, illustrate the confusion that reigns in this area. Two of the misconceptions, that oral agreements are fundamentally different from contracts and that because a contract is "binding," i.e., enforceable, you will very likely use force to execute the agreement, will be dealt with first because they are related and lead to one of the most important reasons for negotiating a contract.

As recently as the 16th century, only agreements involving real property and monetary debts were enforceable in the civil courts. Other agreements, for example a promise to perform a particular act, were adjudicated in the church courts. It was considered a sin to break a plighted, i.e., sworn, bond and the punishment was excommunication. Despite problems with vague language, difficult verification, and failure of the parties to understand the extent to which they were bound by the agreement, a handshake usually sealed the bargain. After all, hardly anyone could read or write, and God was the ultimate witness.

To this day, oral agreements are at least potentially enforceable contracts, and execution of all contracts is enforced more by moral compulsion that by threat of civil lawsuit. Herein lies the first major reason for negotiating a contract, namely, to make it clear to all parties that an agreement with specific terms has in fact been made. After months of casual conversations in which various amounts of money, space, and personnel have been discussed, it will be impossible to recall how definite everyone was without a final recapitulation. Although a preliminary oral statement is fine, the final terms should be in writing.

Why is a *written* contract so necessary? By far the most important reason for demanding a written contract is to flush out intentional fraud. Someone who is trying to mislead you will be most unwilling to make a written commitment knowing that it is potentially enforceable whereas the private promises that preceded it were not. When you encounter strong resistance at this point you are most likely dealing with a liar. If this becomes the sticking point, you are better off spending another year as a postdoctoral fellow than as an assistant professor in this environment. The argument that the institution does not customarily make contracts with faculty should be checked very carefully with emphasis on the adherence of this particular individual to oral agreements. The argument is completely invalid for the institutions with poor faculty-administration relations: 43% of faculty from these schools report that oral agreements are frequently or usually disregarded. The corresponding figure for written agreements at these institutions is 24%: deplorable, but not quite as bad.

Many additional reasons for getting a written contract exist, of which the following are only a few examples. Who can remember their exact words after two hours, much less after months or years? The person with whom you made the agreement may not even be around when it is executed.

Remember, with rare exceptions, you are being hired by an institution not by an individual, and your agreement is with that institution. The individual is only acting as an agent, so the institution is still obligated even if the individual dies or resigns before you arrive. This specific situation will be considered in more detail in Chapter 4. Also, a written contract is more easily renegotiated or enforced. If you are asked to give something up because the budget was cut (a favorite excuse), then what is the other side willing to give up? What about less teaching time or, at least, a promise (written of course) to give first priority to making up the deficit next year. Without a written agreement in the first place, your chances of exacting compensating concessions are nil. Enforcing contractual agreements by means other than moral suasion is a distinct problem to be discussed in Chapter 4.

Why a *detailed* written contract? Simply because a vague contract is equivalent to no contract at all. Unqualified statements such as "laboratory space," even if "X square feet" are specified, are useless without room numbers or similar designations. Chairpersons have been known to walk prospective faculty through suites of "available" space only to put them in one room with two other assistant professors when they arrive. If funds for supplies and equipment are involved, who will be authorized to expend the funds, and precisely what equipment will be purchased. Just as you would set up a laboratory procedure, think your way through each situation and try to anticipate potential problems. Certainly, discuss the process with a senior professor who has already gone through—virtually anyone sensible with whom you are on good terms will do.

MODEL CONTRACTS

The following examples of clauses taken from three different contracts illustrate the foresight and attention to detail required by faculty and chairpersons.

A. Your appointment will be as Assistant Professor in the Department of
———.

B. Your appointment as Assistant Professor in the Department of ———
will be renewable yearly. Future consideration for promotion and tenure will be in accordance with the Policies of the Board of Trustees as they are established from time to time.

C. Your initial appointment as Assistant Professor in the Department of
_____ will be for a term of 3 years, renewable at the end of that term by
mutual consent. Your credentials will be submitted to the CAPQ and a decision on tenure will be reached by the sixth year of employment.

Contracts limiting consideration of appointment and promotion procedures to Clause A, are equivalent to no contract at all. Clause B is better
but has two major problems. Yearly renewal of contracts for tenure-track
faculty is absolutely unacceptable. If administrators want the option of
dumping junior faculty when severe personality problems or inability to
work independently become evident, a three-year term accomplishes that
objective more humanely. The second sentence of Clause B suggests, but
does not state, that the candidate will actually be proposed for promotion
and tenure at the appropriate time. Certain chairpersons are fully aware
that they can avoid liability (see Chapter 3) by allowing a contract to ter-
minate while failing to propose a candidate for tenure. Clause C solves
these problems without unfairly compromising administrative prerogatives.

A. The salary for your position will be $_____ per year.

B. Your total salary will be $ _____ per year and will be comprised of
one or more components related to academic service and sponsored research
activities. Salary levels and the relative composition of the above salary components are determined by the Chairperson and Dean as part of the College's
annual budget process.

C. The salary level for your position is $_____ per year. This component
will initially be payable by the University and subject to increments as stipu-
lated in contracts negotiated by the faculty and staff collective bargaining unit.
It is possible to supplement this basic salary from research grants by up to
20% of the maximum for academic rank (in your case this would amount to
approximately $_____ per year in supplementation funds). I expect that you
will make every effort to put 50% of your salary on externally funded grants
by the end of the initial 3-year appointment.

Clause A is inadequate because it makes no commitment to pay the
salary indicated and does not allow for increments. Even a 5% inflation rate
will increase prices 28% in 5 years. If your faculty collective bargaining
agreement does not incorporate a cost-of-living adjustment then write your
own. Clause B is only marginally better. It still does not provide for increments and, on careful reading, does not commit the school to pay a particular salary. Next year, you could be asked to pay 100% of your salary from
grants if someone makes a mess of the endowment portfolio. Clause C

makes a very firm commitment, but asks for something in return—a reasonable deal for both parties.

A. Your laboratory and office space will be made available in accordance with our previous discussions.

B. Your laboratory and office space will be in Room _____.

C. Initially, the laboratory space that would be available to you is Room _____, as I previously showed to you. I view this initial assignment as temporary, and as renovations proceed in the Department it would be necessary to move you to another laboratory, most probably one of the ones currently occupied by Dr. _____. As I wish to disrupt research activities as little as possible, I would prefer you to remain in Dr. _____ laboratory as long as possible. It is my policy to periodically review research space needs in the Department on the basis of group size, available funding, etc. and I anticipate that as your research efforts grow you will compete favorably for available space.

"You must be joking!" is the most appropriate response to Clause A. Clause B is better provided the department is stable, but would probably require renegotiation in a few years as requirements change. Clause C recognizes that the departmental resources are in flux and protects the interests of both parties to a reasonable degree under the circumstances.

A. As we discussed, start-up funds in the amount of $_____ are available for new faculty in the Department.

B. Start-up funds in the amount of $_____ per year will be available to you for a period of 2 years. These funds may be used for supplies, small items of equipment, and a technician's salary, as you wish.

C. We can offer start-up funds in the amount of $_____. These funds may be accessed over a period of 2–3 years and may be spent on supplies, equipment, salaries, or for whatever purpose you choose. In addition, the Center sets aside research funds on a competitive basis for new and junior faculty in the amount of up to $10,000 per investigator and I am confident that you will be in a promising position to have access to these funds. We will provide one technician for a period of 3 years or until you have secured support via external funding.

Clause A does not commit any of the indicated funds to you. Clause B is simple and direct enough for most situations. Clause C is remarkably similar to Clause B in content but details additional options and conditions so that the offer can be viewed in the most favorable light.

A. (No mention of responsibilities)

B. Your appointment will include your entire professional effort and anticipates your willingness to participate in the educational and research programs of the Department and the College.

C. Junior faculty are not expected to have a significant medical school teaching load during the first year after their employment. It would be beneficial to you and to the Department if you were to participate in graduate-level teaching programs in the areas in which you have expertise. Further, I expect that you will be assigned minimal committee duties and administrative functions during the first 2 years after your arrival. Every effort is made to distribute teaching and administrative responsibilities equitably among the faculty.

Clauses A and B are equivalent—there is the implicit assumption that faculty will assume whatever teaching responsibility they are assigned and discharge this obligation as they see fit. If this system worked perfectly, contracts would be unnecessary. Clause C is preferable since it provides for a phase-in period and stipulates an equitable distribution of responsibilities. Many schools, and faculty as well, might prefer to specify a specific number of student contact hours. This precaution is unnecessary for most graduate faculty since classroom hours are minimal, but is definitely indicated for faculty who might be required to teach undergraduate courses.

Every matter of potential importance should be dealt with similarly. To name a few: relocation expenses, parking space, patient care responsibilities and remuneration, patent rights and royalties, consulting activities, specific fringe benefits, availability of indirect grant costs for faculty use, availability of shared equipment and departmental services, provision for graduate student and postdoctoral fellow stipends, computer facilities, and secretarial support.

Chapter 3

Lawsuits

In hearing litigation, I am not different from any other man. But if you insist on a difference, it is, perhaps, that I try to get the parties not to resort to litigation in the first place. Analects, XII.13.

Thoughts of a lawsuit against school administrators rarely enter the minds of enthusiastic new faculty: they rarely leave the minds of faculty trapped in a negative academic environment. More than 40% of survey respondents from these schools had personal knowledge of colleagues taking this course of action and 75% had heard of lawsuits against the university administration. I know of one school, perhaps an extreme example, where at least ten separate lawsuits were undertaken in as many years. Academics in this environment may rightly consider the lawsuit a normal part of faculty life and a potential remedy for years of frustration.

Of course, the uniquely litigious character of our society also affects the thinking of university faculty. "I'll sue the bastard!" must be one of the most frequently thought phrases in the English language. Not only physical pain or monetary loss resulting from another's actions, but insults, off-color remarks, and any adverse administrative actions seem to offer the possibility of huge financial gain, or at least revenge, by resorting to legal action. In reality, legal remedies are available for very few of the innumerable annoyances plaguing faculty lives and resorting to these remedies is usually an exceedingly bad way to solve the problem.

Academic departments and colleges as social organizations are equivalent to hunting bands or social clubs in that members are expected to conform to the expectations of the group. You don't conform, and membership is withdrawn. Specifically, problems are to be resolved internally, so appealing to an outside agency identifies you as a traitor who has betrayed the group's trust. Even if you are right, if you believe that you are acting in the best interest of the institution and if you win the case, you will be ostracized and irrevocably branded a troublemaker. Although this prospect is sufficient deterrent for most, there are several additional considerations.

IF YOU INTEND TO SUE

Rule 1 Cool off and introspect.
Rule 2 Reject any "nuisance action."
Rule 3 Exhaust *all* alternatives.
Rule 4 Leave (if possible).
Rule 5 Commence action on "contingency" basis.

This handbook assumes that the reader is seeking the solution to a genuine problem, not one that has been intentionally created with the object of seeking legal redress. Faculty wishing to follow the example of the professional accident victim who walks in front of a moving automobile, perhaps by manipulating an adverse tenure decision to give the appearance that it was based on racial discrimination, embody the destructive mentality that this handbook is intended to neutralize. Viewed pragmatically, perhaps there are untapped possibilities for financial gain in such courses of action, but the risk/reward ratio approximates that of suicide.

Proceeding on the assumption of integrity, clear and effective analysis of the situation is essential. If a few minutes are required to regain composure after a minor argument, then it is not unreasonable to reflect for a few days or weeks on a situation that you feel might justify a lawsuit. During this cooling-off period reflect on the role that you played in creation the situation. If you have encountered similar problems in other environments, e.g., at home or in a different institution, then perhaps the difficulty lies in your ability to deal with superiors, inferiors, or peers. Possible solutions range from reading popular books on assertiveness and other social skills to professional counseling and psychotherapy. Above all, talk to friends and colleagues about the problem. If they have encountered similar difficulty with a chairperson or with another colleague, then you will certainly feel more confident of your sanity and you may have established a very useful liaison.

Baseball players know that the umpire will not change the call but they argue anyway just to "keep him honest." Similarly, corporate legal staffs commence some legal actions with no expectation of winning simply to let the competition know that they have their eye on the ball. Private individuals sue their neighbors to get revenge by making trouble. The individuals initiating these legal actions are defining their territory against outsiders and are generally respected for doing so by members of their own group. As already pointed out, legal actions by faculty are generally directed against members of one's own group so the resulting social stigma neutralizes the nuisance value.

Moreover, certain categories of lawsuits must be considered nuisance actions not because of motive but because the legal cause of action is weak or because they simply cannot be won. The courts will not interfere with decisions on promotion and tenure by a Committee on Academic Promotions and Qualifications (CAPQ) or equivalent elected body. The courts defer to the professional expertise of such bodies, and you accept their decisions implicitly when you take the position. Similarly, business decisions made within the bounds of applicable contracts cannot be successfully contested no matter how capricious they might seem. Yes, without a contract, employees can be fired solely because the boss doesn't like them. Slander suits are almost never won because the term's definition requires that a defamatory statement that damages one's reputation or well-being be made with malicious intent. Unless the slanderer openly reveals malicious intent to reputable witnesses, a singularly stupid slip, they can get off by claiming that it was their moral duty to pass on this information. Of course, anything said in private can simply be denied.

Only two remaining causes of action are relevant to academics: violation of statutory civil rights and breach of contract. Whereas CAPQ and other administrative decisions cannot be contested on the basis of professional competence, the claim can be made that they discriminated on the basis or sex, race, religion, or national origin. Be aware that a recent U.S. Supreme Court decision held that state employees were exempt from liability for money damages under Civil Rights Law 1983.[1] Less commonly, the consequences of failure to fulfill contractual obligations are sufficiency damaging to justify a lawsuit.

Regardless of whether your case is potentially winnable, actionable, or merely "fightable," all reasonable solutions to a damage suit must be sought. Not only does honesty dictate that you sincerely try to solve the problem, but the documentation obtained in the course of seeking a solution will be invaluable if a lawsuit is eventually filed. You must appeal in writing and receive a written reply from your chairperson, dean, president, and university chancellor, moving up the administrative ladder as you are rejected at each lower level. You must meet with these officers and discuss the matter if possible. Any appeals permitted within the faculty governance organization, e.g., of CAPQ decisions, must be pursued. The American Association of University Professors (AAUP) should be contacted and any recommended negotiation should be undertaken. Union grievances must be filed and appealed to arbitration. Avoid only administrative procedures that do not allow recourse to civil litigation, such as appeals to the State Human Rights Commission. Union arbitration is to be undertaken if the decision is not binding on the employee. Of course, written documentation must be obtained at every stage. Remember, you are usually much better off with a negotiated compromise than with a protracted battle.

If a compromise cannot be negotiated, your next best course of action is to leave the institution. Your present acute problem aside, faculty are constantly moving on to better positions, and your relocation may be overdue. Unless your current institution has a highly positive atmosphere as defined by this survey, finding a more satisfactory work environment should be possible and doing so has only to be balanced against the psychological strain of relocation. The present problem with its associated psychological, social, and economic costs shifts the balance strongly in favor of departure. Inaction may seem a reasonable alternative, but consider that if compromise on this issue is impossible then it probably won't be the last to plague you. The question of when to leave will be considered at greater length in Chapter 4.

If you can't leave and can't just forget the whole thing then locate a suitable law firm to advise and, possibly, to represent you. If you have arrived at this point, there are only two additional considerations: that the firm have expertise in your area of litigation and that they accept a contingency fee. Family lawyers, friends, and the local Bar Association can recommend firms in your area that have represented faculty. Weed out any that cannot point to specific court victories or satisfactory out-of-court settlements. Although few faculty can afford the cost of bringing a complicated case to trial, willingness to accept a contingency fee, usually one-third of any settlement or court award, is not merely a matter of economics. It is also a stringent test of the merit of your case. If you are willing to pay $250–$350 an hour for representation, wherein lies the incentive for a lawyer to say that you will certainly lose. This is like offering to pay a house painter by the hour and then asking whether your house should be painted. On the other hand, a firm willing to invest its own money in your case must be reasonably certain of eventual profit.

Having embarked on a course of litigation, how long will you have to wait for a resolution? Count on five to seven years not including appeals. Perhaps you feel that this isn't exactly swift administration of justice and perhaps it isn't, but it is miraculous that the system works at all when only one of the four major players is interested in a speedy resolution. Certainly the defendants don't want their actions held up to public scrutiny and possibly censure. The longer they delay the closer they come to retirement. There is also the possibility that you will die or become so discouraged that you drop the case. Their lawyers are being paid by the hour so protracted litigation is a gold mine. The longer the case runs without a settlement the larger your damages become so even your lawyers' fees mount as the case drags on.

Meanwhile you suffer. If conditions were bad before filing the suit they only become worse afterwards. You can no longer use even the implicit threat of legal action as a bargaining chip so all negotiations are halted—frequently until the day before the trial is scheduled. Thus, you are not only subjected to social opprobrium, but also locked for five to seven years into a situation that you previously thought to be intolerable. This is not to discourage anyone from seeking legal redress for genuine wrongs, but rather to make it clear that the lawsuit is a serious undertaking. It is truly an action of last resort because, once commenced, it forecloses any other solution.

SEXUAL HARASSMENT

Sexual harassment may involve the behavior of a person of either sex against a person of the opposite or the same sex. In 1980, the Federal Equal Employment Opportunity Commission issued a definition, which, at the author's institution, has been expanded to include any unwelcome sexual advances, requests for sexual favors, and other verbal or physical conduct of a sexual nature, when:

1. Submission to such conduct is made either explicitly or implicitly a term or condition of an individual's employment or status as a student;
2. Submission to or rejection of such conduct is used as the basis for decisions affecting the employment or academic status of that individual;
3. Such conduct has the purpose or effect of unreasonably interfering with an individual's work performance or educational experience, or

creates an intimidating, hostile, or offensive work environment. A hostile environment is created by, but not limited to, discriminatory intimidation, ridicule, or insult. It need not result in an economic loss to the affected person.

Dealing with unwelcome sexual conduct is much more difficult than defining it, but significant progress has been made in recent years. The following rules are based on published guidelines[2] and are consistent with the approach toward undesirable employment conditions taken throughout this handbook.

DEALING WITH UNWELCOME SEXUAL CONDUCT

Rule 1 Investigate your prospective supervisor.
Rule 2 Reject unwelcome advances.
Rule 3 Prepare for remedial action.

Exposure to sexual harassment is an experience best avoided. A history of prior abuses by a chairman, division director, or other prospective supervisor can be elicited only by questioning former department members since fear and embarrassment effectively preclude present victims from revealing their plight to you. Above all, take allegations of sexual misconduct seriously by giving them proper weight in your employment decision—if it happened to others, it can happen to you.

If an incident does occur, then adhering to established and tested guidelines may help to prevent escalation and minimize retaliation. Unwanted advances must be rejected unambiguously. Say no and say it firmly, without smiling. Speak to your supervisor. If the harasser is your supervisor, speak to his/her supervisor, or to someone he/she would listen to. Keep a diary or log of what is happening to you. Include direct quotes, witnesses, or patterns of the harassment. Have your log witnessed periodically. Save any letters, cards, or notes sent to you. Keep both the log and notes in a secure place, preferably at home. If the behavior continues, then tell the harasser in writing that you object and describe the specific things that offend or upset you. Keep a copy of this letter. If these steps fail, talk to the appropriate institutional official.

Chapter 4

Leaving

A man who serves his king and three times finds his advice rejected and still does not leave the country, is hanging onto his post for the sake of the salary. Even though he says that it is not the salary that attracts him, I won't believe him. Liki, XXXII.

Knowing which position to leave is just as important as knowing which position to take. After all, they are merely two aspects of the same process of career development and personal growth. Although progressing to a new level of accomplishment by moving to a new institution is similar in some respects to graduation, it differs by requiring an act of will and entailing many negative consequences. The strain of relocating your family and the prospect of losing a full year of research time frequently seem to outweigh any possible advantages, especially if you have not yet begun to look for the new position. Knowing when and why to make the move helps to make the decision to do so easier.

```
                        TIMES TO LEAVE

    Best    Before you arrive
    Better  After you receive a new grant or renewal
    Good    At premature career plateau
    O.K.    As an alternative to lawsuit
    So-so   You receive an unsolicited offer
```

Recalling over two decades in which job opportunities in academic research and teaching have been very sparse, professors are understandably reluctant to jeopardize a position that they have already accepted. This attitude would become increasingly inappropriate and counterproductive if the widely predicted shortage of academics, including scientists and engineers, eventually materializes. Even now there are two major reasons for turning down a position that you have already accepted.

You should continue to seek a still better offer, i.e., one at an institution more conducive to career development or one that provides you with more salary and support funds. This is dishonest since you have given your written word that you will come, but the very chairpersons with whom you are dealing have employed the tactic routinely in their own job search and can hardly fault you for doing so. Continue to interview and contact any previous "possibles." Say it straight—you have received another offer, but you would prefer to be there and wonder whether they might reconsider. Perhaps they have failed to fill the position because of the small package or unrealistic requirements and are more interested in you. If you do turn down the first deal, cite "personal reasons" like everyone else.

If the institution breaks your contract before you even arrive or gives you reason to believe that it might do so, then you should renegotiate the entire deal. You should not stop work at your old institution until all essential new space, equipment, and personnel are in place. The option to collect your paycheck without being present full-time, until your laboratory can sustain a corresponding full-time operation, should be written into your contract. If your future chairperson dies or starts job hunting before you arrive, you must get written assurance from the dean that your contract provisions will be honored and you must ask each of the senior faculty their opinion on whether you should come. One of them will be the acting chairperson, possibly until you come up for tenure. If they think you shouldn't come, then don't. Finally, you may have read this handbook, done your investigation, and discovered that you have made a terrible mistake. Not to worry. By now everyone else will have read it too, so no one else will be fool enough to take the offer as currently structured and you have plenty of latitude for renegotiation.

Because not arriving at the new job is a real possibility, you should make your personal plans with this contingency in mind. Set your arrival date for January so that you can delay selling your house and moving your family until the spring or summer as delays emerge. Considering a spouse's career requirements is not the handicap that it might seem. You can both follow that same rules outlined in this handbook and, at least to outsiders, blame any delays on the other party.

Do not resign from your current position no matter what the pressure until one month before you are ready to move, i.e., about the same time that you engage the movers. Do not even reveal to colleagues at your current institution that you are going to move until you formally resign. This is particularly important for tenured faculty—unbelievable animosities surface, sometimes with catastrophic effects, once colleagues learn that you are moving up and they are not. Discretely take care of your subordinates (graduate students, fellows, and technicians), letting them know what will happen if you do leave, i.e., whether you will take them or not and how the transition will be managed.

Even modest levels of grant support greatly enhance the bargaining position of research-oriented faculty. The costs of maintaining an institution continue regardless of who occupies the space. If, over time, your grants generate a million dollars of indirect costs and cover part of your salary, an institution should be willing to pay one-half of that amount to keep you on board.

If your current institution is not willing to negotiate a better deal when you get assured support, then look elsewhere. Put the reason up front— you have the funding and you are looking for an environment where this is appreciated. Delay the start date of your grants for at least six months to show all concerned that you mean business.

The threat of leaving is a very powerful weapon and should be used with discretion. Never threaten to leave unless you already have a firm offer elsewhere—the bluff might be called much to your embarrassment. The widespread knowledge that you are looking around is usually enough to lubricate negotiations. If not, then everyone is happy to see you go and the threat to do so is useless.

The sensation of reaching institutional limits can be similar to that of hitting a brick wall with your head. Failure to attract high-quality graduate students, fellows, and collaborators is the most common, and the most serious, situation in which this frustrating feeling arises. If you have the ability, the ideas, and the money, but can't get the necessary people, then moving is the only option. Students and even fellows pick the institution first and the lab second. Once there, they frequently pick the least demanding lab and the least demanding project within each lab. Brilliant faculty at mediocre institutions with only difficult problems to offer have literally worked themselves to death trying to overcome these obstacles. Don't make the same mistake.

Although leaving has already been recommended as a preferable alternative to lawsuits, the reason that problems arose in the first place may now have become much clearer. Perhaps you should never have taken the position or should have moved long ago. Correct the mistake now with the tactics of this chapter in mind.

Receiving an unsolicited offer is certainly exhilarating, and many faculty at surveyed institutions would immediately leave if they got one. However, to most faculty or potential job candidates an unsolicited offer should be the signal that they may be a desirable item. Now is the time to start looking seriously for a new position, this time taking the school rankings in Appendixes II–IV and your own evaluation of academic environment into consideration.

Chapter 5

Training

A craftsman who wishes to practice his craft must first sharpen his tools.
Analects, XV.10.

Choosing an institution for graduate or postgraduate training is an even greater challenge for students aspiring to academic careers than for professors seeking employment. As a typical 22-year-old college senior, you may be very intelligent, but you have had little experience evaluating aspects of academic atmosphere that still seems rarefied and inaccessible. This chapter of this handbook will show that rules very similar to those developed for professors can be applied to great advantage by students as well.

EVALUATING A GRADUATE PROGRAM

Rule 1 Evaluate the institution.
Rule 2 Evaluate the mentor.
Rule 3 Evaluate the students.
Rule 4 Evaluate the offer.

The challenge of choosing an appropriate institution is not only more difficult for the student than for the professor, it is also more important. Unless the academic atmosphere stimulates a student and promotes professional development, further progress toward an academic career is unlikely. Without sufficient stimulation to make the academic setting interesting and rewarding, few students are motivated to become professors themselves. In any case, without a solid record of success at the graduate and postgraduate levels, no academic job opportunities will be available.

Fortunately, the questions elaborated in Chapter 6 can be used by students as well as by their professors. Simply substitute "students" or "fellows" for "faculty" as appropriate. Perspectives of the two groups are somewhat different. Therefore, students should ask the questions not only of professors, but of other students at the institution as well. Do not limit your inquiry to the "representative" students designated by the program for your benefit. Get a list of all the current students or fellows and talk to as many as possible selected at random. If the opinions of these students differ from those of the designated representatives, ignore the latter completely.

Even though students are somewhat insulated from the administrative concerns and political intrigues that affect faculty working conditions so strongly, this does not amount to complete isolation. Many of the questions that appear irrelevant to students at first sight, assume more significance on close inspection. For example, contractual obligations and administrative competence become the concern of a student who fails to get the promised stipend. Institutional support for faculty research becomes a factor for the student whose project is suspended for lack of funds. Nevertheless, since the scores in Appendix II were based on faculty responses, they should be used as a general guide and should not be compared with the results of personal surveys conduced from the student perspective. The personal survey results can best be used to compare programs within the student's field of interest.

Choosing a graduate program on the basis of an objective analysis of academic atmosphere is certainly preferable to the usual method based on hearsay and general reputation, but it is only the first step of the process. In contrast with undergraduate education, which is primarily didactic, graduate study is an apprenticeship in which cognitive and technical skills are transmitted from a professor to a junior colleague. The resulting close relationship makes the choice of mentor at least as important as the choice of graduate program. In programs that deemphasize formal courses, the thesis advisor may be the only faculty member with whom the student interacts directly, thereby making this choice so critical that it warrants a separate set of rules.

EVALUATING A MENTOR

Rule 1 Evaluate the field of study.
Rule 2 Evaluate recent productivity.
Rule 3 Evaluate student performance.

Horror stories abound in which bright students pick excellent graduate schools only to find that none of the professors are working on appealing projects. The ensuing trauma could have been avoided if the student had carefully evaluated each professor's field of specialization before entering the program. The argument that no one can be expected to make such fine distinctions so early in their career is entirely inapplicable to students planning an *academic* career. Superior students who have a reasonable chance of succeeding in academia can, and routinely do, make such sophisticated decisions.

If the area of specialization is attractive, then proceed to evaluate productivity. From Nobel Laureate to journalist,[1,2] virtually every writer considering academic training agrees that choosing a productive mentor is a crucial step toward future productivity. In the sciences, one wants to see one or two major papers a year in peer-reviewed journals. Competitively awarded grant support is essential to verify that a professor's work is of the highest caliber and that you, the student, will be properly prepared to compete for grant funds in the future. Somewhat different criteria should be used for humanities professors. One wants to see evidence of seminal contributions, at least two monographs that are well received by the academic community, and several articles in highly regarded journals.

Field of study and productivity can be evaluated simultaneously by getting a printout of articles and books published during the last five years. Read the material if you have not already done so and ask whether you would like to be doing something similar in about ten years. If the answer is "No," then forget about working under this professor. Not all faculty dictate the details of a thesis project, but all certainly require considerable continuity between their own work and that done by their students. They will simply refuse to supervise a project in which they are not interested. In many disciplines, such decisions are nearly out of the professors' hands because they will not continue to receive grant support unless they focus on one or two definite problems.

Ability to place students in academic positions is the ultimate test of a mentor either in the sciences or in the humanities. In certain areas of the biological sciences, e.g., cell and molecular biology, this test has become extremely stringent as the number of trained doctoral graduates has outpaced the ability of funding agencies to support their research. Students considering these fields should carefully consider analyses of this problem that have recently appeared in the professional literature.[3]

If your goal is to become a professor yourself, then you need clear evidence that your prospective mentor is capable of furthering this objective.

Keep in mind that most faculty job placements (65%) are through a previous contact, especially the thesis advisor (40%).[4] Get the names of all past graduate students, or at least five, and check their current whereabouts—national directories of college and university faculty make this step relatively easy. Avoid asking professors directly about their ability to place students. Your questions would probably offend an effective, nationally recognized figure in your field and might throw your academic sophistication into doubt. The main caveat is not to be misled by the case of the single successful student who is presented as "typical" of what can happen if you are good enough. If you are good enough to do everything on your own, then why do you need a mentor at all?

Having satisfied the other criteria, prospective graduate students and fellows who are contemplating an academic career should ask only one question about their stipend offer, "Is one being offered?" If no institution, department, or mentor feels that you merit a stipend, then your aspirations are probably unrealistic. Beware of stipend offers in which the institution, mentor, and other students are not satisfactory. Many programs, especially science programs at highly negative institutions, will offer a stipend to any American graduate who is not overtly psychotic. The likelihood of moving on to an academic position from these programs depends entirely on the record of the mentor. The argument against pursuing a humanities program without a stipend is even stronger because the non-academic career options are more limited. Thus, graduate study in the humanities without the stipend offer that confirms substantial talent is a poor risk best left to those for whom a substantial family fortune renders future income from salary irrelevant.

PART II

SUBJECTIVE EVALUATION OF THE ACADEMIC ENVIRONMENT

Chapter 6

U.S. Medical Schools

It is difficult to be a ruler, and it is not easy to be a subject either. Analects, XIII.15.

Other chapters deal with questions that are frequently asked by academic job candidates. This chapter deals with questions that are usually avoided, but must be asked if the characteristic intellectual and social atmosphere of an institution is to be assessed. The approach used in this evaluation is based on a survey of academic working conditions at U.S. medical schools conducted by the author in 1989 and expanded to include major universities, Canadian medical schools, selected liberal arts colleges, and pharmaceutical companies in 1992. Responses to the survey questions are used to construct a picture of atmosphere and working conditions at each institution.[1]

The responses of U.S. medical school faculty are discussed after each question. In order to avoid excessive detail, the responses of other respondent groups, e.g., faculty of major universities and selected liberal arts colleges, are described only if they differed significantly.

The term used for the sum of subjective and objective factors that determine atmosphere and working conditions is termed "academic environment." This term is also appropriate for industrial research organizations in view of the efforts that many companies make to emulate positive attributes of the academic environment. Although the subjective phase of the evaluation aimed to be comprehensive by including many aspects of career development, faculty-administration relations as well as purely emotional factors, it is by no means exhaustive. Any reader can easily identify facets of academic life not covered in the survey. This limitation notwithstanding, enough areas are covered to build up a recognizable picture of conditions at the institutions surveyed.

33

Whether a subjective assessment of academic environment by the faculty or professional staff of an institution accurately reflects the scholarly contributions of their institution is a question to be analyzed in detail, but one that is only indirectly relevant to prospective faculty seeking employment. If staff at some institutions are productive but loathe their working conditions, whereas staff at other institutions are equally productive but satisfied, then the subjective evaluation of academic environment becomes a critical parameter by which desirable and undesirable positions can be distinguished.

EVALUATING OBJECTIVES

Objective 1 To provide an objective system for comparing academic environment within a group of institutions.

Objective 2 To identify institutions having a highly negative environment so that they can be avoided.

The evaluation can be conducted casually. In fact, the questions need not be asked in the form suggested or even directly. Simply try to remember key words or phrases as they arise in the course of normal conversation and ask questions whenever they seem appropriate. Later, note the number of different people who gave each response. Thus, if two different people used the word "enthusiastic" in describing the atmosphere of the school, then "enthusiastic" gets two ticks in the "Yes" column. If one person says "enthusiastic" and the other says "not enthusiastic," then this word gets one tick in the "Yes" column and one tick in the "No" column. The word gets only one tick no matter how many times the same person uses it. Ignore neutral responses such as "somewhat," "borderline," and "no difference."

You might expect some of the key words and questions to have absurdly predictable responses. Is not every academic institution intellectually stimulating? Are not hard work and productivity highly valued at all universities? Not only is the answer to both these questions "No" but they are retained in the evaluation scheme precisely because the survey response was found to be highly correlated with institutional atmosphere. In fact, the response to each of the questions that follows was found to differ significantly ($P < .01$), and in many cases to differ dramatically, between institutions with highly positive versus highly negative atmosphere. To avoid unnecessary repetition, the evaluation questions will be discussed using the terms "negative school" and "positive school" when referring to institutions having highly negative and highly positive environments, respectively.

Keep in mind that at the end of this exercise the number of ticks in the "positive" and "negative" columns will be summed to arrive at a score that can be used objectively to distinguish positive environments from negative environments. Clear answer patterns emerge after talking to as few as three individuals, so even readers not inclined to computation can make informed judgments once they start asking the right questions.

ACADEMIC ENVIRONMENT
EVALUATION QUESTIONS

	Positive	Negative
1. Will locating at this institution help me move to a better position in the future?	Yes	No

For the university professor, just as for the corporate manager or scientist, large gains in status, income, and resources are frequently made by moving to a new institution. Survey respondents from positive schools were four times more likely to answer "Yes" to this question than those from negative schools. Check the "No" column for this question only if the answer indicates that your chances of moving up will be reduced by locating at this institution. Forty-seven percent of survey respondents from negative schools gave a "No" response—a very strong argument against taking such a position.

	Positive	Negative
2. Will taking this position enhance my ability to build a research group and obtain funding?	Yes	No
3. Will taking this position enhance my ability to develop productive collaborations and do good work?	Yes	No

These two questions explore some of the factors that underlie the answer to Question 1. Survey respondents from positive schools were two times more likely to answer "yes" to this group of questions than respondents from negative schools. Conversely, the latter group was three times more likely to answer "no." Several differences in the responses to individual questions were very pronounced since only 13% and 10% of respondents from positive schools gave negative answers to Question 2 and Question 3, respectively.

To the extent that these questions evaluate basic requirements for success in academic research, faculty at positive schools feel that their chances of becoming successful have been enhanced. Faculty at negative schools are much less likely to make this judgment. Although colleges and universities are not pure meritocracies, failure to achieve some degree of success in research virtually precludes any move to a better position at another institution. As worded, the questions are strictly applicable only to laboratory-based scientists, but they can easily be adapted to evaluate positions in the social sciences and even in the humanities.

4.	Is the general atmosphere stimulating?	Yes	No
5.	Is the general atmosphere optimistic?	Yes	No
6.	Is the general atmosphere competitive?	Yes	No
7.	Is the general atmosphere enthusiastic?	Yes	No
8.	Is the general atmosphere apathetic?	No	Yes
9.	Is the general atmosphere pessimistic?	No	Yes
10.	Is the general atmosphere paranoid?	No	Yes
11.	Is the general atmosphere crazy?	No	Yes
12.	Is the general atmosphere desperate?	No	Yes

Observing the usage of these nine key words is a powerful technique for distinguishing positive from negative environments. Survey respondents from positive schools cited the terms "stimulating," "optimistic," and "enthusiastic" three to seven times more frequently than those from negative schools. Conversely, respondents from negative schools were equally likely to cite the terms "apathetic," "pessimistic," "paranoid," "crazy," and "desperate" when referring to their institutional atmosphere. The resulting pictures contrast vividly. Although the questionnaire was not specifically designed to evaluate abnormal psychosocial factors, the responses suggesting higher frequencies of overt psychopathology at negative schools were consistent with other responses suggesting widespread disregard of accepted standards of behavior (see below).

13.	Is grant support highly valued?	Yes	No
14.	Are publications highly valued?	Yes	No
15.	Is productivity highly valued?	Yes	No

16. Is teaching highly valued?   Yes   No

Faculty from positive schools felt that contributions in all these categories were highly valued. For teaching, the ratio of positive to negative responses was 5:1 and, for the other questions, this ratio was in excess of 10:1. Because even faculty at negative schools tend to feel that these standards of academic productivity are valued, the ratio of positive to negative responses is most indicative of institutional atmosphere. Thus, the corresponding ratios for negative schools were only 1:1 for teaching and 2:1 for the other questions.

17. Do hardworking and productive faculty receive proper recognition and reward within the institution?   Yes   No

The response pattern for this question was very similar to that for all the questions related to career development. At positive institutions, positive responses exceeded negative responses by a ratio of about 10:1 whereas at negative institutions the corresponding ratio was about 1:1. In fact, only 34% of respondents from negative institutions gave an affirmative answer to this question. Answers to such questions could not be correlated with the productivity of individual respondents because the survey was anonymous. However, it seems reasonable that the more highly productive the individual, the more distasteful an atmosphere in which productivity is disregarded.

Readers who are wondering whether academics value, recognize, and reward qualities besides academic productivity have anticipated the next set of questions.

18. Is competence highly valued?   Yes   No
19. Is reliability highly valued?   Yes   No
20. Is sense of humor highly valued?   Yes   No
21. Is erudition highly valued?   Yes   No
22. Is clique membership highly valued?   No   Yes
23. Is cunning highly valued?   No   Yes
24. Is arrogance highly valued?   No   Yes
25. Is fame highly valued?   Yes   No

Clique membership and cunning appear to be highly valued at negative schools because each of these terms was cited by about one-fourth of respondents from these institutions and much more rarely by respondents from positive schools. Conversely, erudition and sense of humor were each cited by less than 10% of respondents from negative schools, but by from 29% to 68% of respondents from positive schools. The picture at positive schools was as expected for an orderly, ethical meritocracy. The picture at negative schools where cunning and erudition were cited with equal frequency was equally clear.

26.	Is competition mainly constructive?	Yes	No
27.	Is competition mainly fair?	Yes	No
28.	Is competition mainly helpful?	Yes	No
29.	Is competition mainly vigorous?	Yes	No
30.	Is competition mainly backbiting?	No	Yes
31.	Is competition mainly discouraging?	No	Yes
32.	Is competition mainly destructive?	No	Yes
33.	Is competition mainly insidious?	No	Yes
34.	Is competition mainly exciting?	Yes	No
35.	Is competition mainly tiring?	No	Yes

Competition is viewed as constructive, fair, or helpful by 43% to 68% of respondents from positive schools, but by less than 25% of respondents from negative schools. About 20% of the latter group used terms such as "discouraging" or "destructive," whereas only 5% of respondents from positive schools used these terms. Fewer than 8% of the group from negative schools felt that competition was exciting or vigorous, whereas 20% to 30% of the respondents from positive schools characterized competition in these terms.

This set of questions reveals a characteristic of positive schools that might not be viewed as pleasant by all prospective faculty. They are not quiet, benign, or "laid-back" institutions. Competition is at least as intense as that at neutral or negative schools and probably more so given the close correlation between institutional productivity and vigorous competition discussed in Chapter 9.

Positive schools are distinguished by the style with which this game is played. If competition is vigorous, fair, and constructive within a highly stimulating intellectual environment (see Question 39), then a successful competitor dominates by force and clarity of intellect. At negative schools where competition is described in diametrically opposed terms, successful competitors rely on cunning and clique membership in order to achieve dominant positions. Prospective faculty should carefully consider their personal life strategies before immersing themselves in a potentially alien environment. Behavior eliciting respect at a positive school would probably lead to ostracism at a negative school. Conversely, a perfectly adequate intellectual energy level for a negative school would probably be considered moribund at a positive school. Since virtually no one is happy at a negative school, perhaps individuals who feel that they are best suited for this environment should pursue a nonacademic career path.

36. Is the physical plant adequate?	Yes	No
37. Are faculty free of minor irritations?	Yes	No
38. Have faculty been victims of major crimes?	Yes	No
39. Is intellectual stimulation adequate?	Yes	No
40. National rank of school top 25%	bottom 25%	

Faculty from negative schools were four to eight times more likely to respond "No" to the first three questions in this set than those from positive schools and 38 times more likely to respond that intellectual stimulation was inadequate. Although 56% of faculty from positive schools felt that their respective institutions were in the top 25% nationally, 9% of faculty from negative schools also gave that response. Similarly, while 38% of faculty from negative schools felt that their institutions were in the top 25% nationally, 10% of respondents from positive institutions gave that response. This finding reinforces the notion that perception of national rank guarantees neither a highly positive nor a highly negative academic atmosphere.

41. Is general faculty welfare given adequate consideration?	Yes	No
42. Is minority faculty welfare given adequate consideration?	Yes	No
43. Are faculty involved in administration?	Yes	No
44. Are the dean and president respected for their academic accomplishments?	Yes	No
45. Is it generally known that faculty have filed lawsuits against the administration?	No	Yes

Faculty at negative schools are as dissatisfied with the institutional administration as with any other aspect of their environment. They respond "No" to the general faculty welfare question over four times as frequently as respondents from positive schools. They respond "No" to the minority welfare question over twice as frequently. Fully 40% feel that their dean or president are not respected as academicians compared with only 15% at positive schools.

Faculty at positive schools report that they are involved in administration twice as frequently as faculty at negative schools. Although only 18% of faculty from positive schools reported direct knowledge of lawsuits by colleagues against the institutional administration, 44% of faculty from negative schools reported that they were aware of such actions. Whereas not one respondent from a positive school reported filing such a lawsuit, four out of five of the lawsuits reported in this survey were at negative institutions.

46.	Is overall administrative competence at least adequate?	Yes	No
47.	Are faculty-faculty and faculty-chairperson disputes usually resolved amicably?	Yes	No
48.	Are administrative decisions rational?	Yes	No
49.	Is the administration "top-heavy?"	No	Yes
50.	Is faculty research adequately supported?	Yes	No

A similar pattern emerges when more detailed perceptions of administrative competence are elicited. Faculty from negative schools are five times more likely to feel that overall administrative competence is inadequate. They are three times more likely to feel faculty-administration disputes are not resolved amicably, that administrative decisions are irrational, that the administration is top-heavy, i.e., overstaffed, and that faculty research is not adequately supported.

51.	Are written agreements frequently broken?	No	Yes
52.	Are oral agreements frequently broken?	No	Yes

If anyone could still be wondering whether faculty-administration relations actually affect them directly, then these two questions are for you. Twenty-four percent of faculty from negative schools report that written agreements are frequently broken, and 43% feel that oral agreements are frequently broken. These frequencies are six and three times higher, respectively, than reported by faculty at positive schools. Although such figures cannot be translated into an exact probability that your contract will be broken, they are symptomatic of the disregard for generally accepted standards of conduct that prevails at negative schools.

53. Do faculty from within the institution ever interact
    socially?   Yes   No

Whereas 16% of faculty at positive schools report that social interactions with colleagues from within the institution (dinner at home, parties, etc.) almost never occur, 35% of faculty at negative schools make this statement. Thus, the increased potential for social isolation at a negative school must be considered.

Score one positive point for each advantage mentioned and one negative for each "No."

54. Are there advantages of locating at this institution
    that outweigh all the disadvantages?   Yes   No

Examples used in the survey were: high salary, hard money, intellectual atmosphere, easy to raise a family, convenience, geographic location, unique equipment (facilities or resources), collaborators, spouse's career, and recreation. Strictly comparable scoring requires counting only these terms. Common sense dictates that the terms should apply to you, at least in principle. Faculty from positive and negative schools tend to cite the same terms. However, those from positive schools cite all advantages with greater frequency (in the case of intellectual atmosphere and collaborators, with much greater frequency).

55. Would you leave to take a position of equal rank
    and salary elsewhere?   No   Yes

This question is the "bottom line" because it evaluates the net result of disadvantages and compensating advantages. The question is designed to eliminate effects of salary, and the survey responses do not, in fact, show a statistically significant correlation with salary. Forty-two percent of faculty from negative schools, but only 12% of faculty from positive schools responded "Yes" to this question. From the fact that nearly one-half of faculty at negative schools have not left, one must infer that they cannot leave without making an unacceptable sacrifice.

SCORING THE ACADEMIC ENVIRONMENT EVALUATION

Step 1 Add up responses in the positive column and responses in the negative column separately.

Step 2 Divide the total number of positive responses (P) by the total number of negative responses (N).

Step 3 A resulting P/N ratio greater than 4.20 is highly positive, from 4.20 to 1.60 is neutral, less than 1.60 is highly negative.

Examples of scores produced by this evaluation system are given in Appendix II. Both the absolute values and the range of the results are striking. More than 90% of the scores are positive, suggesting that the positive aspects of most academic environments generally outweigh the negative aspects. This result is encouraging even though the questions are not designed to yield a result of 1.00 for a neutral environment (whatever that may be). However, the highest and lowest scores differ by a factor of more than 100 with practically no negative responses at one end of the spectrum and a 3:1 negative to positive ratio at the other.

The boundaries defining highly negative and highly positive academic environments isolate approximately 25% to 30% of schools at either end of the response spectrum. Of course, scores of 1.59 and 1.60 are not significantly different—they merely define the boundaries of groups that are significantly different from one another ($\chi^2 > 6.6$). Therefore, the most practical use for the numbers is to aid in ranking the institutions that have made employment offers and in eliminating those falling at the extreme lower end of the spectrum.

FAMILY LIFE EVALUATION QUESTIONS

The series of questions dealing with personal life and family matters is included as a separate section of the evaluation. These considerations are important. About 90% of survey respondents were married and had children. For these faculty, quality of primary or secondary schools and time spent commuting become critical parts of any compensation package. Moreover, virtually no one willingly exposes themselves to crime and everyone needs a place to live during the 108 hours remaining after a 60-hour work week.

On the other hand, except for the question on major crime, the responses to this set of questions did not differ significantly between highly positive and highly negative schools. Factor analysis showed that even though the responses to the family life questions were not highly correlated with any questions in the academic atmosphere section of the survey (again, except for the question on major crime),[2] the questions formed an independent group with responses that were highly correlated internally. Therefore, the questions are evaluating a set of parameters that differs from school to school, but are unrelated either to productivity or to the subjective aspects of academic environment discussed so far. In this section of the evaluation, the terms "highly positive" and "highly negative" refer to ratings with regard to the family life questions exclusively.

		Positive	Negative
F1.	Will locating at this school make raising a family relatively easy?	Yes	No

Stress the "relatively"—no variation in location can make raising a family easy—and assume that if the answer is not a straightforward "Yes," then they mean "No." Fifty-two percent of survey respondents answered "Yes" and this fraction did not differ significantly when respondents were grouped by sex, marital status, or having had children. However, 86% of respondents from the schools rated most highly positive with regard to family life responded "Yes" while only 12% of respondents from the most highly negative schools gave this response. The following questions examine some of the reasons that these responses were given.

		Positive	Negative
F2.	Have faculty been victims of major crimes?	No	Yes
F3.	Have faculty been victims of minor crimes?	No	Yes

As for Question 38 above, major crime is defined as murder, rape, or mugging. At the most highly positive schools, fewer than 1% of respondents indicated any level of victimization or even that they frequently heard of such incidents. At the most highly negative schools, 41% reported a high level of awareness, i.e., that they had direct knowledge of victimization or at least frequently heard of such incidents. Minor crime, defined as theft, auto vandalization, etc., was much more prevalent, with 51% of total respondents reporting a high level of awareness, and 31% citing direct knowledge of incidents. At the highly positive schools, 15% reported a high level of awareness while, at the highly negative schools, 86% gave this response.

Location of the crimes was not specified, but the correlation with both academic atmosphere and family life question sets suggests that respondents had both school and home environments in mind.

F4. Is commuting time from home to school 30 min-
   utes or less?   Yes   No
F5. Does commuting significantly interfere with
   work?   No   Yes
F6. Is above average child care available and afford-
   able?   Yes   No

Every parent with an employed spouse knows that a family with young children absolutely precludes long commuting, defined here as a commute of more than 30 minutes one way. Fewer than 1% of respondents from highly positive schools, but fully 42% of respondents from highly negative schools, were burdened with long commuting times. Only 11% of respondents from highly negative schools felt that commuting significantly interfered with work, but none (not even one) of the respondents from highly positive schools found that this was so. All but one (96%) of the respondents who reported that commuting significantly interfered with work was married, and all but 4 (88%) had children. Perhaps these faculty find working at home more difficult.

Arranging for daytime child care is a universal facet of academic family life except for the few college and university faculty still living in an extended family. Of the faculty responding to this question, 74% of those from positive schools felt that availability and affordability of child care was above average, and only 4% felt that it was poor or very poor. At negative schools 40% felt that child care was above average; 19% said it was poor or very poor. The second figure is large enough to make careful scrutiny of the local situation imperative.

F7. Are schools above average?   Yes   No
F8. Is above average housing available and affordable?   Yes   No

Because of the present level of the average public school, everyone should demand above average schooling for their children. At highly positive schools, most respondents (81%) felt that they were already getting it, and nearly all the remainder felt that the schools where they lived were of average quality. Even most respondents from highly negative schools (59%)

felt that their schools were above average but, more important, 23% felt that the schools were poor or very poor.

Answers to the housing and schools questions were similar. At highly positive schools, 92% of respondents felt that their housing was above average, and fewer than 1% felt that it was below average. At highly negative schools, 61% felt that housing was above average, while 20% felt that it was below average. The impact of school and housing availability on quality of family life depends on individual satisfaction with "average" living conditions. If college and university faculty generally expect these essentials to be of above average quality, then those at highly negative schools are two to four times more likely to be dissatisfied than those at highly positive schools.

The relationship of commuting to housing and school quality was also examined. Perceived quality of housing did not depend on commuting time for total respondents. However, respondents from highly negative schools were significantly more likely to judge their housing to be above average if they commuted more than 30 minutes (79% versus 54%). All respondents were more likely to judge schools to be above average if they commuted more than 30 minutes (82% versus 72%), but this was particularly true of respondents from negative schools (83% versus 41%). These findings confirm the common sense interpretation that faculty commute long distances to take advantage of superior housing or public schools and that faculty at negative schools are compelled to do so by the lack of available facilities in proximity to work.

SCORING THE FAMILY LIFE EVALUATION

Step 1 Add up responses in the positive column and responses in the negative column separately.

Step 2 Divide the total number of positive responses (P) by the total number of negative responses (N).

Step 3 A resulting P/N ratio of greater than 14.5 is highly positive, between 14.5 and 1.76 is neutral, and less than 1.76 is highly negative.

Examples of scores produced by this evaluation system are given in Appendix II. As for the academic atmosphere evaluation, highly positive and highly negative schools were defined on the basis of scores that differed significantly ($\chi^2 > 6.6$) from the mean family life score of 4.87. Just as for the academic atmosphere section, the range of scores is very large with the highest and lowest having scores with a greater than 80-fold difference. Highly positive schools were characterized by large numbers of positive responses and virtually no negative responses. Only 13 of the 123 medical schools surveyed were rated as highly positive in both the academic atmosphere and family life categories. They were: University of Alabama, Loma Linda University, Chicago Medical School, University of Iowa, University of Kentucky, Michigan State University, University of Minnesota (Duluth), University of Mississippi, University of Rochester, University of North Carolina, Oregon Health Sciences University, University of Virginia, and Marshall University. If anyone is surprised to find certain schools on this list, perhaps their highly positive qualities simply deserve more recognition.

Several schools with highly positive academic environments were rated highly negative with regard to family life. These were: University of California at Los Angeles, University of California at San Francisco, Johns Hopkins University, and Mount Sinai School of Medicine. The outstanding academic reputations of these schools may make coping with their undesirable urban locations tolerable. Conversely, several schools with highly negative academic environments were rated highly positive with regard to family life. These were: Southern Illinois University, Albany Medical College, Medical College of Ohio, Wright State University, and West Virginia University. Especially constructive department chairpersons can make positions at such schools extremely attractive.

Neutral schools had a wide range of scores if the total number of positive and negative responses was small, i.e., if most responses were "average." Many of these schools were not defined as highly positive because they did satisfy the stringent statistical requirement for inclusion in that category. They should be carefully considered by any prospective faculty member, especially if the academic atmosphere score is highly positive or neutral.

Most highly negative schools had scores greater than 1.00. This can be taken to mean that even the schools least conducive to family life have some positive attributes. However, the argument that no one needs more trouble seems more compelling. Faculty already struggling to do good work, get grants, and take care of children will not benefit from the additional burdens of commuting, poor housing, and below average schools. Avoid these problems if you can.

Chapter 7

Alternatives

It is only the most intelligent and the most stupid who are not susceptible to change. Analects, XVII.3.

Alternatives to the traditional high profile research centers have become attractive to prospective faculty in both relative and absolute terms. As funding for faculty full time equivalents (FTEs) has remained static or decreased at U.S. medical schools, many research-oriented universities have emerged as leaders in the industrial and agricultural applications of biotechnology. Certain liberal arts colleges previously considered to be, at best, quaint bastions of medieval thought with heavy teaching loads, now embrace a variety of sophisticated research programs in an appealing atmosphere. In parallel with these developments, the recent emergence of molecular genetics as an academic discipline with commercial applications has opened pathways between academic and industrial careers that in the past were relatively limited. Many postdoctoral fellowships are offered by biotechnical/pharmaceutical companies and major universities are encouraging applied research programs. With continued blurring of the distinction between academic versus industrial positions, movement from one to the other and back again has become feasible.

Equally dramatic has been the emergence of significant research programs within expanding community hospitals at the fringes of major metropolitan areas. The author's present institution, Staten Island University Hospital, is a noteworthy example. Driven by competition for house staff and the most highly trained medical/surgical specialists, these facilities offer academically oriented physicians the opportunity to carry out research of limited scope. Although these centers have emerged too recently to display much of a track record, they appear to offer an attractive niche for certain individuals.

These career alternatives (except for community hospitals) have been examined using the same questionnaire format that was applied to medical schools. The results are presented in detail below and should be encouraging to anyone wishing to consider all of the available alternatives.

RESEARCH UNIVERSITIES AND INSTITUTES

Taken as a whole, responses to the survey questionnaire by faculty at this group of institutions were virtually identical to those by faculty at the group of U.S. medical schools initially surveyed. Thus, the survey data were pooled, and the same definitions of positive, neutral, and negative academic environments based on P/N score were applied to both groups of institutions. These criteria should be applied with the same caveats as before, namely, that even though highly positive and highly negative groups of institutions are clearly different, schools ranking close to one another are probably not even though they fall in different groups.

Despite the overall similarity in faculty assessments, responses by faculty at research universities and institutes to a few individual questions were significantly different. More held nonacademic, non-tenure track positions and salaries were higher with a larger percentage in the greater than $75,000 per year categories. Percentage of hard money and the fraction devoting more than 50% of their time to teaching were also increased. Responses to questions probing the extent to which institutional goals were perceived and accomplishments recognized were more positive and clear-cut. Respondents were more likely to feel that hard work was rewarded, that institutional research support was good or excellent, and that their career advancement possibilities were enhanced.

These faculty were more likely to feel that the atmosphere was stimulating and less likely to feel that the spirit of competition was helpful. A much higher proportion of respondents felt that department chairpersons, deans, etc. were competent and that administrative intervention was successful in resolving disputes. University faculty reported much more direct involvement in administrative decision making. The majority of respondents (61%) ranked their institution in the top 25% of their peer group—not a surprising assessment in view of the manner in which the survey group was selected. Such differences notwithstanding, about the same fraction (27% versus 26% in U.S. medical schools) said that they definitely or most likely would move to a position of equal salary and rank elsewhere.

In the social realm, consulting income and family wealth were valued more highly at the medical schools whereas, at the universities, fame was more highly valued. The spectrum of responses on schools and housing by university faculty was significantly narrowed with about three-fourths reporting "good" or "adequate" to these questions. On the other hand, over four-fifths of university faculty reported "superb" or "excellent" recreational facilities and greatly reduced direct exposure to both major and minor crime. These differences are probably due to the semirural location of many universities in contrast with the more urban location of medical schools.

The cumulative effect of such response patterns was that none of the major research universities was rated highly positive with respect to academic environment and highly negative with respect to family life. Ten institutions were rated highly positive in both the academic atmosphere and family life: University of Connecticut, University of Georgia, University of Illinois, University of Massachusetts, University of Nebraska, Princeton University, Cornell University, Carnegie Mellon University, Pennsylvania State University, and Washington State University.

LIBERAL ARTS COLLEGES

The following 50 schools have been identified as the best small colleges for science in the United States.[1] The criteria included: the percentage of freshmen scoring above 600 on mathematics portion the SAT, the number of NSF fellowship winners in the preceding 5 years, and the number of doctorates in science awarded to graduates from 1920 through 1980.

Despite the excellent academic reputations of the schools in this group, relatively few faculty (only 1 or 2 in many schools) were both sufficiently accomplished and sufficiently inclined to obtain FASEB membership. Therefore, rather than using so few responses to characterize entire schools, the data were pooled and considered to represent these institutions as a group.

Albion College, Albion, MI
Alma College, Alma, MI
Amherst College, Amherst, MA
Antioch University, Yellow Springs, OH
Barnard College, New York, NY
Bates College, Lewiston, ME

Beloit College, Beloit, WI
Bowdoin College, Brunswick, ME
Bryn Mawr College, Bryn Mawr, PA
Bucknell University, Lewisburg, PA
Carleton College, Northfield, MN
Colgate University, Hamilton, NY
College of the Holy Cross, Worcester, MA
College of Wooster, Wooster, OH
Colorado College, Colorado Springs, CO
Davidson College, Davidson, NC
Denison University, Granville, OH
DePauw University, Greencastle, IN
Earlham College, Richmond, IN
Franklin and Marshall College, Lancaster, PA
Grinnell College, Grinnell, IA
Hamilton College, Clinton, NY
Hampton University, Hampton, VA
Harvey Mudd College, Claremont, CA
Haverford College, Haverford, PA
Hope College, Holland, MI
Kalamazoo College, Kalamazoo, MI
Kenyon College, Gambier, OH
Lafayette College, Easton, PA
Macalester College, St. Paul, MN
Manhattan College, Riverdale, NY
Middlebury College, Middlebury, VT
Mount Holyoke College, South Hadley, MA
Oberlin College, Oberlin, OH
Occidental College, Los Angeles, CA
Ohio Wesleyan University, Delaware, OH
Pomona College, Claremont, CA
Reed College, Portland, OR
St. Olaf College, Northfield, MN
Smith College, Northampton, MA
Swarthmore College, Swarthmore, PA
Trinity College, Hartford, CT
Union College, Schenectady, NY
Vassar College, Poughkeepsie, NY
Wabash College, Crawfordsville, IN

Wellesley College, Wellesley, MA
Wesleyan University, Middletown, CT
Wheaton College, Wheaton, IL
Whitman College, Walla Walla, WA
Williams College, Williamstown, MA

The total P/N scores of 3.80, 17.13, and 0.72 for academic atmosphere, personal life, and productivity, respectively, yield a pattern unique to this group of schools. Overall atmosphere is toward the high end of the neutral group whereas the productivity subset is near the midpoint of the spectrum. The personal life and family response is highly positive. The responses to individual questions help to reveal the details of this pattern.

The demographics of the respondent group was similar to those from U.S. medical schools and universities, except that the faculty were overwhelmingly nonminority. Although one-half of respondents reported incomes less than $50,000 per year, the lower salary scale was felt to afford a lifestyle comparable to that of their peers. Nearly one-half did no research despite their credentials, and three-fourths devoted more than 50% of their time to teaching. Research and publications were reported to be valued to the same degree as in medical schools, but 95% of respondents from liberal arts schools felt that teaching was highly valued. More (42% versus 17%) also felt that their contributions were definitely recognized. As for research universities above, the selection criteria resulted in 87% of respondents placing their institution in the top 25% of their peer group. The fraction who would definitely or most likely leave to take a comparable position elsewhere (32% versus 25%) was not significantly different from that at medical schools.

Institutional characteristics cited significantly more frequently were intellectual atmosphere, unique equipment, and ease of raising a family (the last much more frequently). Interactions were less frequently characterized as competitive or paranoid and more frequently as crazy or homelike. Grant support and consulting income were less likely to be valued whereas religion, compatibility, erudition, and sense of humor were cited more frequently. Administrative relationships were like those in major universities rather than medical schools in that department chairs were characterized as competent, disputes were resolved amicably, and faculty were almost all (97%) directly involved in administrative decisions. Both written and oral agreements were rarely broken.

PHARMACEUTICAL COMPANIES

This set of institutions also exhibited a unique profile with P/N scores of 3.40, 45.33, and 1.54 for academic atmosphere, personal life, and productivity, respectively. The respondent demographics were very similar to those of medical school faculty except that nearly twice as many (41% versus 21%) had been at their present institution for less than five years and, of course, none were tenured. None reported incomes less than $50,000 per year and most (62%) were in the $51–$75,000 per year range. Nearly all (90%) felt that this salary provided a lifestyle comparable to that of their peers. Also, nearly all (90%) worked a 40–60 hour week consisting of mostly research (86%), no teaching (73%), and some consulting (44%). Many more (65% versus 28% at medical schools) felt that the present position equipped them for future career advancement, but essentially the same fraction (28% versus 25%) said that they would leave to take a comparable position.

Certain institutional characteristics were rated very highly on the quantitative scales and significantly higher than medical schools. For example, research support was nearly always considered good or adequate (93%), physical plant was at least good (90%), and intellectual stimulation at least good (68%). Administrative parameters were rated similarly to universities and liberal arts colleges with good administrative competence, regard for staff welfare, and lack of top-heavy administration predominating. Awareness of lawsuits was sharply lower (7% versus 29% in medical schools), but other administrative parameters were similar to medical schools.

Subjective assessments differed from those of educational institutions in a number of response categories. Advantages of locating at the institution cited less frequently were recreation and easy promotion whereas high salary, hard money, and unique equipment were cited more frequently. The atmosphere was described as stimulating and competitive whereas the spirit of competition itself was less likely to be characterized as tiring and more likely to be characterized as exciting and political. Grant support and consulting income were not highly valued, but dress, reliability, compatibility, competence, erudition, clique membership, and business sense were reported to be highly valued.

The P/N score for personal life and family reflected almost complete lack of negative responses. For example, no respondents reported intimate exposure to major crime and only 3% to minor crime. This result tends to validate the survey approach because corporate planners are known to have selected locations for their research facilities that are the most desirable from every conceivable standpoint.

CANADIAN MEDICAL SCHOOLS

Data were collected on this group of institutions not because American scientists seriously consider the Canadian job market in their career planning, but because Canadians have historically sought training and employment in the United States and might gain some insight from the survey approach. In addition, Americans can see one version of a system the has evolved under what has come to be known here as "managed care."

The spectrum of P/N scores for academic atmosphere is similar to that the U.S. counterparts minus the most highly positive and highly negative extremes. Thus even though McGill University qualifies as highly positive, it ranks 27th in the survey. Similarly, University of Saskatchewan is considered highly negative, but it is 23 places above the lowest ranked institution in the survey overall. Nevertheless, two schools (Queen's University and University of Alberta) were ranked highly positive with respect to both academic atmosphere and family life.

Respondent demographics show that the faculty are more likely to be tenured and to have been at the present institution for more than 20 years. Only 7% reported salaries less than $50,000 and the remainder were nearly evenly divided above and below $75,000 per year. Fewer devoted more than 75% of their time to research and many more devoted a small proportion of their time to clinical work, i.e., 1%–20%. Although clinical activities were much less highly valued, only one-half as many respondents as at U.S. schools felt that hardworking, productive faculty would fail to receive proper recognition.

Responses to key words were generally similar to those of faculty at U.S. medical schools except that consultant income, clique membership, and family tree were cited significantly less frequently. Competition was less likely to be considered helpful (14%). In contrast, responses to questions on administrative interactions paralleled those from faculty at research universities. Thus, 69% felt that concern for faculty welfare was good or adequate and 70%–80% gave that assessment of administrative competence. Disputes were usually settled amicably, but only after administrative intervention, and 96% reported that they were involved to some degree in administrative decisions.

All of the Canadian medical schools are ranked around or above the middle of the survey from the standpoint of personal life and family and most of the P/N scores are highly positive. The most dramatic example of the data underlying this result is seen in responses to the question on schools which none of the respondents considered less that adequate and which

20% considered superb. Only 1% rated housing less than adequate, and 94% responded superb or good to this question. Recreation was rated less than adequate by only 5% of respondents and superb or good by 87%. Although relatively few cited location as an advantage of locating at the present institution, this appears to result from comparisons among equally desirable locations.

PART III

OBJECTIVE EVALUATION OF THE ACADEMIC ENVIRONMENT

Chapter 8

Productivity

The Master said of Yen Yuan, "I watched him making progress, but I did not see him realize his capacity to the full. What a pity!" Analects, IX.21.

College and university faculty are concerned, even obsessed, with the effects of institutional affiliation on their intellectual productivity. The tools of their trade may range from intellect alone to multibillion dollar equipment, but all have chosen this career path expecting to create something of value. Many base their self-esteem entirely on realizing this goal. Academics also know that tenure and ultimate prestige depend on making worthwhile contributions in their field.

Research grants and publications are the most widely accepted measures of productivity.[1] Research grants recognize past contributions as well as good ideas and plans for future work. The process of writing a fundable research grant is a major creative undertaking and it is recognized as such by peers. Research grant dollars are a measure of research value because, in nonprofit institutions, the value of the product is equal to its cost of production. Books, research papers, and journal articles are merely the media through which insights and research findings are communicated. Highly critical screening by editors and journal reviewers insures that each publication has some original ideas or data and has at least one minimum publishable unit of information. Therefore, the number of publications rather than their actual content or impact can be used as a crude measure of productivity.

THE RULE

Intellectual productivity depends on the individual faculty member—not on the institution.

How can this be so? Everyone knows that some institutions are "better" than others. In every field, prominent faculty seem to be concentrated in a handful of departments. Schools of all sorts have even been ranked on a numerical scale and placed in classifications ranging from "distinguished" to "adequate plus."[2] Appendices II–IV show that grant support and number of publications vary greatly among institutions. Given the abundance of both subjective and objective criteria, prospective faculty assume that their productivity will depend on the institution in which they work.

The assumption is false—at least for the vast majority of faculty. My conclusion is based on objective analysis of grant support and publication rates at a series of institutions, a study that requires some explanation. I believe that the analysis can be applied to both academic and nonacademic institutions because the methods that were initially applied to U.S. medical schools have now been extended to major universities and research institutes. However, academic institutions were used to generate the data, so complete generality cannot be strictly proven.

EVALUATING INSTITUTIONAL RESEARCH GRANT SUPPORT

How are the variations in grant support among institutions to be explained if the commonly invoked, but vague, explanations based on quality, excellence, or political clout are rejected? Failure to resolve this problem has given credence to the generally accepted myths and was partly responsible for motivating the survey of institutional atmosphere already described. Initial attempts to explain productivity in terms of academic environment also failed because P/N score for academic atmosphere at U.S. medical schools was only weakly correlated ($r^2 = 0.29$) with research grants or publications expressed on an institutional basis or per capita, i.e., per full-time faculty member. The general conclusion that overall academic environment does not determine productivity leaves the job-hunting professor free to choose an optimal academic and administrative atmosphere without compromising career development.

A more detailed analysis of the individual survey questions most strongly correlated with grant support and publication rate is undertaken in the next chapter. However, the finding that per capita productivity is roughly constant over a wide range of academic and administrative environments consistent with generally accepted personnel management theory. Productivity of a self-motivated college or university professor whose job itself provides intrinsic rewards would not be greatly influenced by environment.[3] Even if motivation could be increased, much of the academic environment evaluation deals with attitudes of faculty toward their institution, so it is not necessarily true that a favorable attitude toward the school will result in increased productivity.[4]

Searching for parameters that might explain variations in grant support revealed a strong correlation with total research journal publications and a weaker correlation with total number of faculty.[5] Explaining grant support in terms of publication rate is like explaining number of chickens in terms of total eggs laid. Which comes first? Because grant support is necessary for the extremely expensive modern biomedical research that generates the data for publications, publication rate was rejected as an explanation for grant support.

Further analysis of the weak correlation with total faculty showed that the level of institutional grant support could be explained almost entirely on the basis of the number of faculty who were members of the Federation of Societies for Experimental Biology (FASEB).[6] Number of FASEB members, designated F, accounted for 81% and 86% of the variance in dollar amount and number, respectively, of 1993 National Institutes of Health (NIH) grants.[7] Regression analysis showed that for U.S. medical schools:

$$\text{NIH grants (million \$)} = 0.59\ F - 33$$
$$\text{No. of NIH grants} = 2.12\ F - 97$$

For all U.S. universities, medical schools, and research institutes surveyed the following relationships accounted for 78% and 85% of the corresponding variances:

$$\text{NIH grants (million \$)} = 0.53\ F - 22$$
$$\text{No. of NIH grants} = 1.93\ F - 65$$

The coefficients in these equations have remained essentially constant since 1988 allowing for inflation and overall increased numbers of FASEB members. Nevertheless, mere correlations must be interpreted with care.

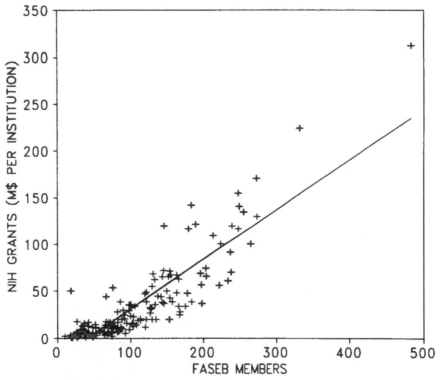

Figure 8.1 Relationship between the number of FASEB faculty members and the total dollar amount of NIH grant support per institution in 1993. Points represent the data for 181 U.S. medical schools, universities, and research institutes.

NIH grant awards are not restricted to FASEB members, nor do staff at all institutions have the same likelihood of joining an FASEB society. Moreover, there is substantial statistical uncertainty associated with each of the numerical coefficients. These caveats notwithstanding, even casual inspection of Figure 8.1 and Figure 8.2 confirms the remarkable relationship between number of FASEB members and grant productivity.

These relationships are difficult to appreciate intuitively because, without looking at individual curricula vitae, real ability is difficult to distinguish from adept social posturing. Even within the schools themselves, faculty often do not know the specific qualifications of their colleagues. Certainly, many faculty with no experience or credentials whatever claim the capacity to do grant-worthy academic research—if only they had the time. Heavier reliance on credentials, as in this analysis, should reduce the confusion.

Why are grant support and FASEB membership correlated? Society membership characterizes a class of basic science–oriented faculty qualified

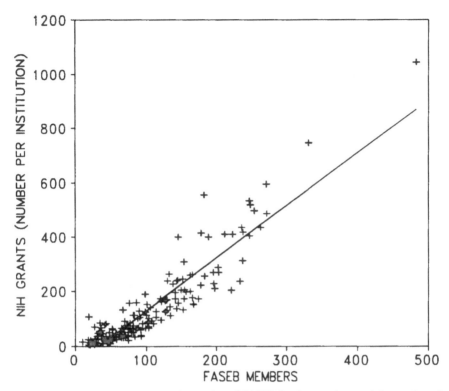

Figure 8.2 Relationship between the number of FASEB faculty members and the number of NIH grants per institution in 1993. Points represent the data for 181 U.S. medical schools, universities, and research institutes.

to do research. Election to the FASEB societies requires evidence that the candidate is an established scientist with a record of original publications; therefore, the fact that FASEB members and other like them get grants is not surprising. It is surprising that the number of well-qualified faculty is the only important parameter determining the level of institutional grant support.

This is not to say that the probability of a FASEB member having a grant is the same at all institutions. Solving the above equations for number of FASEB faculty shows that at zero grants the number of faculty is 46 and 56 for U.S. medical schools and 34 and 42 for all relevant academic institutions. Every institution has faculty, including FASEB members, who are not active researchers. If we tentatively assume that they do teaching, administration, or patient care, then the value 34–56 approximates the average number of FASEB faculty engaged in such activities. Thus, a new faculty member who locates at a school with 40 FASEB members and is required to perform the same teaching and administrative duties as everyone else, has approximately zero probability of getting a research grant. On the other hand, any faculty member who is not among the 30–45 support staff, but is instead devoting nearly full-time to research, has the same probability of getting a research grant at this institution as at any other.

Looking at Figures 8.1 and 8.2, several institutions have exceptionally high ratios of grants per FASEB member (i.e., their data points lie far above the regression line). Those ranking in the top ten with regard to both number of grants and M$ per FASEB member are: Fred Hutchinson Cancer Center, Harvard University, Johns Hopkins University, St. Jude Children's Research Hospital, Stanford University, University of Michigan, and University of Pittsburgh. In several cases, this ranking can be attributed to exceptional selectivity. However, in view of the above discussion, lack of teaching and administrative functions for FASEB members clearly plays a role.

EVALUATING INSTITUTIONAL PUBLICATION RATES

This phase of the project sought to determine the rate at which publications were being produced by each of the institutions for which academic atmosphere scores had been obtained. Publication counts were obtained directly from the National Library of Medicine electronic database rather than by laboriously searching the most frequently cited biomedical journals as was done for the 1988 edition of this handbook. This method yielded a greater than six-fold increase in the number of papers from U.S. medical schools relative to the 1988 study and more closely approximates the total number of publications from each institution. Appendix III lists institutional sources for the 92,050 papers identified in this study.

The number of publications per institution varied from 0 to 2944. Analysis of these data (Figure 8.3) showed that, just as for NIH grants, institutional publication rate was most closely correlated with the number of FASEB faculty. This correlation accounted for 87% of the variance in publication rate; including additional parameters (e.g., P/N ratio, total faculty, or number of students) did not improve the correlation or account for additional variance.[8]

The following equation describes the relationship between publications and FASEB members (F):

$$\text{No. of Publications} = 5.2\,F - 108$$

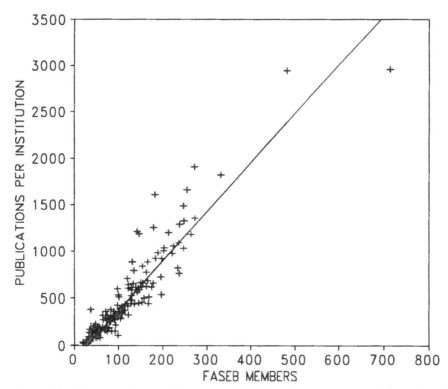

Figure 8.3 Relationship between the number of FASEB faculty members and the institutional publication rate. Data points represent total number of publications during 1993 in biomedical journals for 212 U.S. medical schools, universities, and research institutes.

Keeping in mind the usual caveats, solving this equation for zero publications also reveals 21 FASEB faculty per institution who did not publish in 1993. Undoubtedly because the much larger number of journals included, the number of nonpublishing faculty is about one-half the estimate of 35–40 obtained in 1988. The close correlation between institutional grant support and publication rate suggests that it is the same faculty who neither get grants nor publish. Just as for grant support, faculty members who are *not* in this group have nearly the same publication rate regardless of their institutional affiliation.

CRITICAL MASS

The concept of "critical mass" can be reassessed in the light of this analysis of institutional grant support and publication rate. It is widely believed that some minimum number of scientists must be brought together before their interactions can stimulate maximum output of ideas and work. Such a requirement could explain the low per capita productivity of the schools with small numbers of FASEB faculty. The idea is also intuitively appealing because obviously two scientists interact more than one and because articulating the hypothesis conspicuously displays some knowledge of atomic physics. According to this hypothesis, the 30–40 "nonproductive" faculty represent the subcritical mass that must be accumulated in order to achieve optimal interaction intensity.

Several arguments suggest that this interpretation is incorrect. The analogy with atomic fission dictates that productivity of all group members should increase once the critical threshold is achieved. If this were so, then analysis of institutions with large numbers of FASEB faculty should show few or zero non-producers. Eliminating institutions with fewer than 100 FASEB faculty yields the three relationships shown below:

$$\text{NIH grants (million \$)} = 0.72\,F - 59$$
$$\text{No. of NIH grants} = 2.4\,F - 160$$
$$\text{No. of Publications} = 5.1\,F - 84$$

Solving these equations for zero productivity shows that large schools, in fact, have nearly the same or larger numbers of nonproductive faculty than the group as a whole, i.e., from 16 to 81 in the various expressions. The coefficients themselves are not changed significantly. Thus, increasing mass, i.e., number of FASEB faculty, does not eliminate the constant number of FASEB members who are not getting grants or writing papers.

Moreover, we all know who these people are and what they are doing. For example, at my own institution, I know virtually every FASEB member. Although a few are truly nonproductive, most are doing full-time teaching, administration, or patient care. Increased or decreased opportunities for interaction with active scientists is not likely to alter the area in which they make their contribution, nor should it.

The nearly constant coefficients in the above expressions show that productivity of even the active scientists is not greatly affected by interaction intensity. The possibility that face-to-face interactions are of primarily social value and do not affect productivity at all merits further study. Perhaps BITNET and the fax machine can satisfy the communication needs of any scientist who feels isolated.

A second line of argument leading to the same conclusion is based on second-order regression analysis of the data on institutional grant support and publications rates. In certain systems, e.g., sedimentation in the analytical ultracentrifuge,[9] interactions leading to aggregation can be detected by the appearance of second- and higher-order terms when regression lines are fit to the data. Thus, in the equation

$$y = ax^2 + bx + c$$

the second-order term, ax^2, would be negligible if the postulated interactions did not occur. Temporarily putting aside the question whether such a second-order term would have any distinct meaning in the present context, the data were fit with the following second-order regression equations:

$$\text{NIH grants (million \$)} = 0.0011 \; F^2 + 0.19 \; F - 3.9$$
$$\text{No. of NIH grants} = 0.0027 \; F^2 + 1.07 \; F - 1.9$$
$$\text{No. of Publications} = 0.0013 \; F^2 + 5.8 \; F - 149$$

This modification did not account for additional variance in grant support or publications ($r^2 = 0.84, 0.88$, and 0.87 for grant amounts in dollars, numbers of grants, and publications, respectively). Moreover, the coefficients for all the second-order terms are so small that for the average institution with 97 FASEB members, the ratios of the first- and second-order terms is about 2:1, 4:1, and 44:1 for the above parameters. To the extent that the terms reflect a nonlinear effect of institutional size, no phenomenon of this sort appears to play a significant role in determining total production of grants and scientific publications. As the exact meaning of F^2 is irrelevant in this context, this problem will be dealt with in the next chapter where it assumes some importance.

Why then does scale appear to affect productivity? The most obvious effect of increasing scale is to increase visibility. Even if per capita productivity is the same at two institutions, twenty times more papers will emanate from the one that is twenty times larger. The academic community may not even be aware that the smaller institution exists. Increasing the number of productive faculty also reduces the visibility of the group concerned with primarily internal matters such as teaching and administration. Thus, the proportion of productive faculty increases rapidly as institutional size increases from very small to moderately large. This increase in per capita productivity results from passing over a threshold that is real but unrelated to any requirement for a critical mass of interacting faculty.

Chapter 9

Productivity Versus Atmosphere

Raise the straight over the crooked. This can make the crooked straight.
Analects, XII.22.

Prospective faculty members have legitimate reason to ask whether even further analysis can improve their ability to find a suitable position. Consider the progress made so far: Faculty can now quantify academic atmosphere, evaluate factors that affect family living, and realize that their total production of grants and publications is not greatly affected by institutional affiliation. These are useful tools for evaluating job offers. If further progress in understanding the factors that affect academic atmosphere can be accomplished by analysis, faculty may not be better able to find a position, but they will be much better equipped to recognize a good one when they do find it.

Two general questions will be answered. "How is total institutional production of grants and publications related to academic atmosphere?" and "Can aspects of academic atmosphere be used to predict institutional productivity?" Because we have already shown that total institutional productivity is closely related to the number of FASEB members, these questions aim to identify specific qualities of the academic atmosphere at institutions that are both large and productive.

PRODUCTIVITY IN RELATION TO
ACADEMIC ATMOSPHERE

Data for U.S. medical schools were used, and the analysis was performed in stages. The mean response for each survey question was calculated for

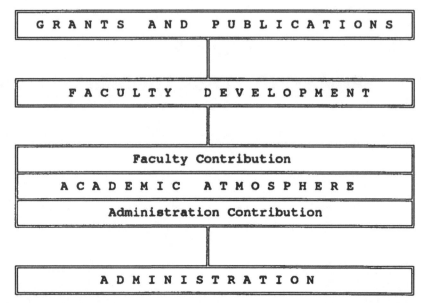

Figure 9.1 Model relating categories of survey responses to productivity measured as total grants and publications.

each institution. Then correlation coefficients were calculated for all pairs of survey questions and for pairings between the survey questions pertaining to grant productivity and total publications for each institution. All pairings with correlation coefficients greater than 0.399 were subjected to principal component or "factor" analysis.[1]

The procedure identified a set of survey responses that was highly correlated with total productivity and formed a discrete factor group. The survey questions in this group dealt mostly with faculty career development, but included several questions probing subjective aspects of academic atmosphere. A second important set of survey responses was also identified. This set was not directly correlated with total productivity, but was highly correlated with subjective aspects of academic atmosphere. The factor group contained mostly questions dealing with administrative characteristics and included many of the same aspects of academic atmosphere contained in the first factor group.[2]

Figure 9.1 shows the relationships among these groups of questions. Survey questions in connected boxes are both highly correlated and form a defined factor group. The subjective aspects of academic atmosphere are placed in a separate box to convey the idea that administrative actions and faculty development overlap in this area alone.

Figure 9.2 details the survey questions forming the factor group closely correlated with total production of grants and publications. Because there are only 16 items in the list, it is clear why overall academic atmosphere defined by 55 questions was not highly correlated with productivity—the relevant correlations were obscured by other often opposing relationships.

The underlying questions have already been discussed as determinants of academic atmosphere. The novelty of the relationship to productivity lies in the link between two completely independent data sets. Faculty evaluated their schools with no reference whatever to the actual level of productivity. In fact, no objective measures of total institutional productivity had been devised at the time the survey was conducted. Finding any factor patterns that make sense under these conditions argues strongly for the validity of the approach.

The pattern does make sense. Faculty at schools with large numbers of grants and publications feel that their career development has been greatly enhanced. The standards of the schools are high and competition is keen, but viewed in a positive light. The atmosphere is stimulating, and they would move only if offered a better position. The correlations imply a converse as well: Faculty at schools with low levels of total institutional productivity give exactly the opposite responses to these questions. Faculty whose primary goal is to maximize professional productivity and who want to be evaluated by strict standards on this basis do not need a survey to evaluate schools. They need only look up the historical level of institutional productivity in Appendices III and IV and weigh the offers from the schools ranked highest. Others who are not so sure can simply consider that responses to this subset of survey questions is likely to have the greatest bearing on their future productivity. Figure 9.3 details the factor group containing survey responses related to administrative actions. Although none of the responses to questions probing administrative competence and style was directly related to institutional productivity, all such responses listed in the lower block of the figure were highly correlated internally and all were highly correlated with one or more responses to questions evaluating subjective aspects of academic atmosphere. Taken together, the two groups of questions formed a distinct factor pattern.

The questions evaluating subjective aspects of academic atmosphere in Figures 9.2 and 9.3 are very similar, suggesting that this is the common ground on which faculty and administration can stand. However, certain differences are apparent. For example, the intensity of competition is correlated with faculty development, not with administrative style, suggesting that the faculty who answered the questionnaire feel that this aspect of

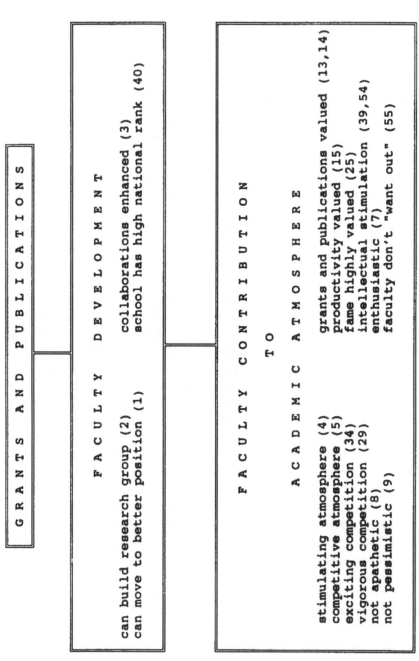

GRANTS AND PUBLICATIONS

FACULTY DEVELOPMENT

can build research group (2)
can move to better position (1)

collaborations enhanced (3)
school has high national rank (40)

FACULTY CONTRIBUTION

TO

ACADEMIC ATMOSPHERE

stimulating atmosphere (4)
competitive atmosphere (5)
exciting competition (34)
vigorous competition (29)
not apathetic (8)
not pessimistic (9)

grants and publications valued (13,14)
productivity valued (15)
fame highly valued (25)
intellectual stimulation (39,54)
enthusiastic (7)
faculty don't "want out" (55)

Figure 9.2 Survey questions relating faculty contribution to academic atmosphere with faculty development. Survey questions (numbers) in connected boxes are both highly correlated and form a defined factor group.

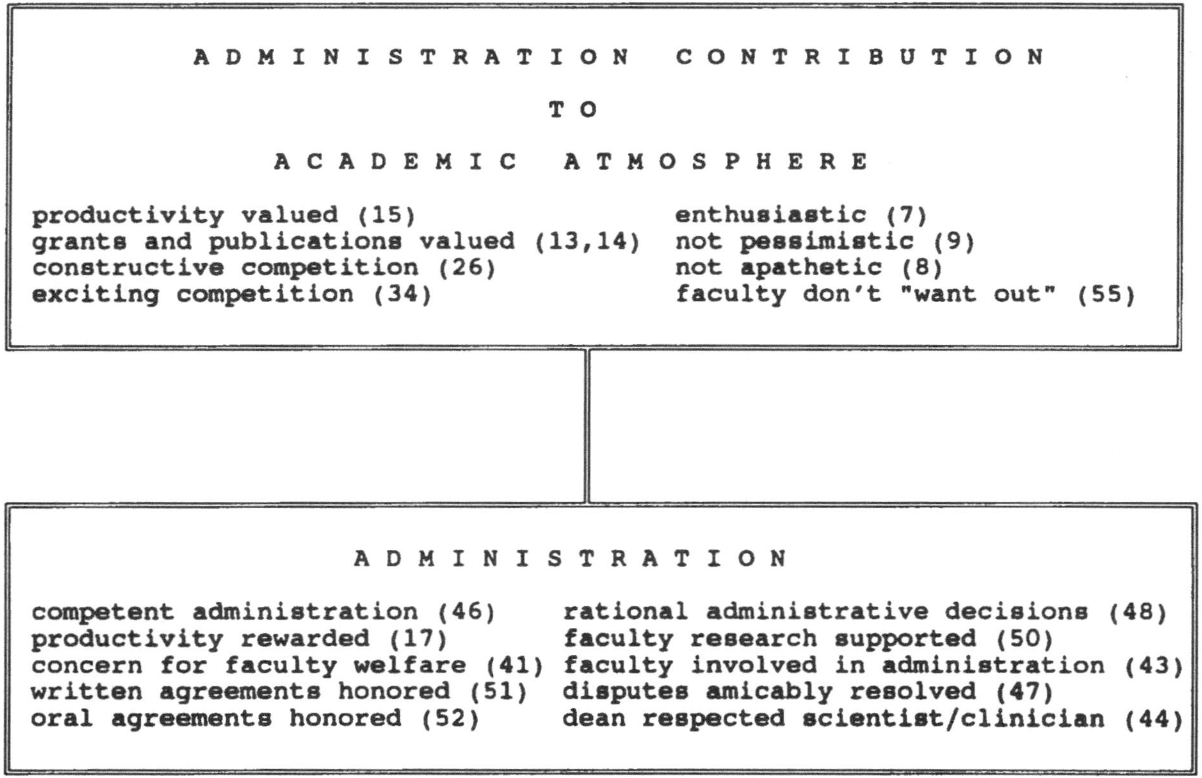

ADMINISTRATION CONTRIBUTION

TO

ACADEMIC ATMOSPHERE

productivity valued (15)   enthusiastic (7)
grants and publications valued (13,14)   not pessimistic (9)
constructive competition (26)   not apathetic (8)
exciting competition (34)   faculty don't "want out" (55)

ADMINISTRATION

competent administration (46)   rational administrative decisions (48)
productivity rewarded (17)   faculty research supported (50)
concern for faculty welfare (41)   faculty involved in administration (43)
written agreements honored (51)   disputes amicably resolved (47)
oral agreements honored (52)   dean respected scientist/clinician (44)

Figure 9.3 Survey questions relating administrative component of academic atmosphere to institutional administration. Survey questions (numbers) in connected boxes are both highly correlated and form a defined factor group.

academic atmosphere depends primarily on them. The constructive nature of competition at highly productive schools is correlated with questions related to the administrative competence at these schools. This suggests a key role for administration in directing the competitive energy of faculty along productive lines.

Questions related to intellectual stimulation are conspicuously absent as are any related to factors that might provide direct motivation, e.g., fame. Thus, faculty appear to feel that administrative factors have little direct influence on intellectual climate except to contribute to an atmosphere that is enthusiastic and that supports their values. Indirect influence is not necessarily weak—the reason most frequently given by faculty leaving an unacceptable university teaching position (35%) was incompetent administration and administrators.[3] The specific questions included in the set describing administrative style and competence give a clear indication of the means by which college and university administrators can have positive effects on academic atmosphere.

Highlighting specific administrative functions in this way allows the role of college and university administration to be viewed analytically. The results are useful to faculty trying to understand how administrators might affect the productivity, and hence the reputation, of their department or school. For example, it becomes clear that because any such effects are indirect, a competent and sympathetic chairperson can insulate faculty from many potential problems. The results are also useful to administrators seeking to increase institutional productivity. Affecting academic atmosphere is the route by which this can be accomplished.[4] If competence and dishonesty are the fundamental problems, these can be dealt with best by trustees.

PREDICTING PRODUCTIVITY

At first sight, it might seem that faculty job applicants have no reason to predict institutional productivity. Individual productivity is independent of institutional affiliation, and the actual current grants or publications are listed in Appendices II and IV. However, there are two good reasons for doing so. Many candidates may wish to consider positions at institutions that were not surveyed or for which current productivity data are difficult to obtain. A convenient means of making this evaluation should be available. Moreover, since production of grants and publications must lag faculty recruitment, a measure of academic atmosphere that is correlated with current productivity of a large number of schools should also effectively predict *future* productivity of institutions that are in transition. Prospective faculty certainly want to know whether the school making them an offer is on the way up or on the way down.

The subset of survey questions most highly correlated with production of grants and publications by medical school faculty is listed separately in Appendix I and can be used to make this evaluation. The resulting Productivity P/N score is reasonably well-correlated with the various measures of productivity introduced so far: total publications ($r^2 = 0.57$), dollar value of NIH grants ($r^2 = 0.64$), and number of NIH grants ($r^2 = 0.61$). These correlations apply to medical schools only—the heterogeneity of nonmedical institutions precluded analysis by this approach. The less than perfect correlations for even the group of medical schools are expected because the subset of survey questions is small and because schools in transition are included.

Identifying medical schools in transition is a major application of the productivity P/N score. Thus, a school ranking very high by this measure, but exhibiting relatively low current productivity appears to be poised for increasing prominence. Conversely, a school with high current productivity, but a low P/N score may well be in decline. Several interesting examples of this application will be given in the next chapter where the correlation between productivity P/N score and production of highly cited publications is discussed.

Chapter 10

Quality

Is a man not superior who, without anticipating attempts at deception or presuming acts of bad faith, is nevertheless, the first to be aware of such behavior? Analects, XIV.31.

The preceding analysis of relationships among publication rate, grant support, and number of FASEB faculty intentionally focused on quantity rather than quality in order to make the results relevant to most college and university faculty. This approach does not disparage the work of the "average" professor in any way but rather diminishes emphasis on factors that might apply to a particular group. Because the majority is excluded by any definition of "elite" and because any such definition is necessarily subjective, avoiding the issue of quality early on was justified. Having shown that two quantitative measures of productivity, grants, and publications are simple functions of institutional size, we now focus on the question, "Is institutional size related to the *quality* of work being done?"

Defining the boundary between elite and average work may be a subjective process, but once defined the most important contributions in any field can be identified quite objectively. In the biomedical sciences, prizes and awards, e.g., Nobel prizes, are not suitable criteria because it is often difficult to identify the institution where the work was actually done and useless to analyze such a small number of cases. Naturally, an opinion poll of sufficient breadth and depth would not be feasible. If "important" papers are defined as those attracting the most attention and forming the basis for the most subsequent work, then the number of times a particular paper was cited by subsequent authors becomes a measure of its importance.[1]

Short lists of most highly cited papers, e.g., the 100 most highly cited, tend to be dominated by descriptions of methods that become widely adopted and therefore highly cited. This bias is not a significant shortcoming because it can be overcome in many ways. In the present project, the lists were extended to over 500 papers so that methods descriptions no longer dominated. In addition, several sets of most highly cited papers were selected from journals that do not publish methods papers, and other sets were selected on the basis of high citation frequency in the year following publication—methods papers are rarely cited shortly after publication because widespread adoption of the method usually takes several years. In all, 805 highly cited papers were attributed to the institutions listed in the Appendices. Any remaining bias toward methods descriptions does not compromise the ultimate objective, namely, to identify some institutions where the most important work in biomedical sciences is being done. Laboratories where widely adopted methods are developed are usually highly productive in more fundamental research as well. Examples abound in which the method development presaged an important breakthrough by the very same laboratory.

Figure 10.1 shows the relationship between institutional size, expressed in number of FASEB faculty, and production of highly cited papers. In contrast with data for total publications shown in Figure 8.3, the data for highly cited publications was fitted quite poorly by a simple linear regression ($r^2 = 0.20$). Fit was greatly improved ($r^2 = 0.77$) by a second-order regression with the following result:

$$\text{Highly cited publications} = 0.00026 \ F^2 - 0.0029 - 0.016$$

This relationship has several striking characteristics. The absolute magnitude of the first-order term is negligible. At $F = 117$ (the average number of FASEB members per institution surveyed) the ratio of first- to second-order terms is about 1:10. This contrasts with the grants or total publications data for which the second-order term could be neglected. The calculated intercept for zero highly cited publications is 5.5, a value not significantly different from zero. Because the second-order term dominates the relationship, the zero intercept suggests that virtually all FASEB members capable of producing highly cited publications are doing so. This interpretation is consistent with the generalization that productivity is largely independent of institutional affiliation.

The precise meaning of the F^2 term assumes more importance than in the previous analysis of total publications where it was negligible. If we recognize that the variable F in the first-order term already accounts for simple interactions that affect the dependent variable, highly cited publications, then the F in the second-order term must have an additional attribute.

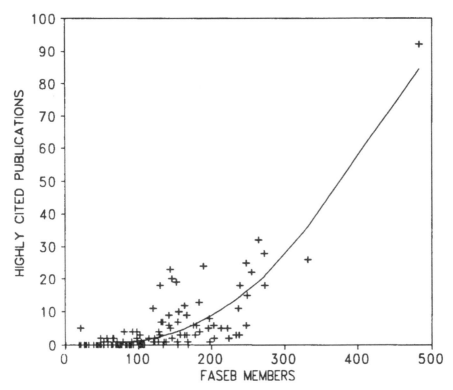

Figure 10.1 Relationship between production of highly cited papers and number of FASEB members. Points represent data for 123 U.S. medical schools.

In analogy with ultracentrifugation, this attribute is incorporation into a dimer. In the present example, this attribute might be incorporation into a research group so that production of an important paper becomes possible whereas it would otherwise have been impossible. This interpretation is unlikely to be correct if virtually all FASEB members capable of producing highly cited publications are already doing so.

Alternatively, the FASEB members in the F^2 term might differ *ab initio* in some critical way from those in the first-order term. For example, they might be more competent and innovative. In that case, they are more likely to be found in large institutions, because these centers are more attractive to them by virtue of opportunity and prestige.

Both alternatives may play a role in determining the production of highly cited papers by large institutions. Faculty heading highly innovative laboratories are usually singularly intelligent and creative individuals who could be productive in virtually any environment. However, being intelligent, they are attracted to schools where they are most likely to get the top graduate students, fellows, and junior faculty to join their research group. These associates may not get all the ideas, but they certainly facilitate development and implementation of the good ones.

A list of the 20 schools producing the largest number of highly cited papers is essentially a list of the great research universities, at least those that have associated medical schools. Thus, visibility by virtue of both sheer size *and* production of important work are probably responsible for the reputations that these schools enjoy. Now the obvious question becomes "Which determinants of academic atmosphere are correlated with production of highly cited publications?"

Regression analysis showed that the following survey questions were correlated with production of highly cited publications: #4 (stimulating atmosphere), #7 (enthusiastic atmosphere), #13 (grant support highly valued), #14 (publications highly valued), #17 (productivity rewarded), #25 (fame highly valued), #40 (school has high national rank), #47 (disputes amicably resolved), #48 (rational administrative decisions). Institutional administrators wishing to attract faculty capable of producing highly cited publications should pay special attention to these determinants of academic atmosphere.

The productivity P/N score discussed in the previous chapter may also be used to assess future production of highly cited publications. All but two of the institutions with more than ten highly cited publications also had productivity P/N scores greater than 1.00. The exceptions were University of Colorado School of Medicine and New York University School of Medicine. Both institutions have an illustrious record of producing important work but have recently undergone financial and administrative turmoil. The atmosphere reflects current problems and predicts decline in the absence of an appropriate response.

Only four medical schools had productivity P/N scores greater than 1.00 but fewer than ten highly cited publications. These were University of Alabama School of Medicine, Duke University School of Medicine, Vanderbilt University School of Medicine, and University of Texas Southwestern Medical Center. Several of these institutions are known to have undergone major expansion recently, and the P/N score suggests that a highly positive atmosphere has helped to make this possible by attracting qualified faculty. The number of highly cited publications should increase over time to reflect this transition. Similarly, many research universities and institutes had productivity P/N scores greater than 1.00. This finding may presage a shift away from the major medical centers as sites of the most important work in biomedical science.

Chapter 11

Conclusions

College and university faculty expect to make valuable contributions and have a good deal of fun in the process. Choice of institution affects both the type of contribution they are most likely to make and how much they will enjoy making it. Using medical schools as a model, faculty at small schools are more likely to be engaged in teaching, administration, or patient care. Faculty at larger schools are more likely to get grants and publish. Nevertheless, those faculty who are getting grants and publishing do so with about equal productivity regardless of institutional size. Thus, individual capabilities and the terms of employment are critical determinants of productivity while the generally accepted reputation of an institution is not.

Academic environment may not affect intellectual productivity, but it certainly does affect a professor's daily life. Faculty at highly positive institutions feel that their contributions are more highly valued and their environment is more pleasant in every respect. Faculty at highly negative institutions are doing the same work, but enjoying it less. Given a choice between happy and unhappy, I recommend the former.

The opinion of faculty that locating at certain schools enhances their career development can be integrated with the objective finding that per capita production of research grants and total publications is constant from school to school for similarly qualified faculty. Attitudes toward research productivity depend on school size, i.e., schools with large numbers of faculty with the qualification of FASEB membership tend to be characterized by more positive attitudes toward research grants and publications. Even without quantitative data on academic environment, institutional cultures that recognize and reward research can be expected to appear especially attractive to faculty qualified to do it. Since FASEB members are qualified to carry on independent research, they are attracted to these schools that increase in size over time. This process of aggregation increases the amount of institutional research activity without greatly affecting individual accomplishment. The reputation of these schools attracts eminent faculty who do the most important, i.e., highly cited, work and further enhance the reputation of the school.

There is a message here for college and university administrators hoping to increase institutional productivity. Attract qualified faculty by improving the academic environment, especially those aspects highly correlated with production of grants and publications. The analytical approach taken here and the intuitive approach of successful corporate management lead to similar conclusions. Both not only attest to the effectiveness of this strategy but also concur that influencing institutional culture is virtually the only route by which administrative goals can be achieved. Faculty would certainly agree as well.

Appendix I

Academic Environment Survey Forms

The forms on the following pages reproduce the questions discussed in Chapter 6. Remember key words or phrases as they arise in the course of normal conversation and ask questions whenever they seem appropriate. Later, note the number of different people who gave each response. Thus, if two different people used the word "enthusiastic" in describing the atmosphere of the school, then "enthusiastic" gets two ticks in the "Yes" column. If one person says "enthusiastic," and the other says "not enthusiastic," then this word gets one tick in the "Yes" column and one tick in the "No" column. The word gets only one tick no matter how many times the same person uses it. Ignore neutral responses such as "somewhat," "borderline," and "no difference."

Score your survey by adding up the total responses in the positive column and the total in the negative column. Divide total positive by total negative and evaluate the result by comparing with P/N scores for the schools in Appendix II or by placing in the following broad categories: P/N for overall academic atmosphere greater than 3.40 is highly positive, between 3.39 and 1.7 is neutral, and less than 1.7 is highly negative.

ACADEMIC ENVIRONMENT QUESTIONS

	Positive	Negative
1. Will locating at this institution help me move to a better position in the future?	Yes	No
2. Will taking this position enhance my ability to build a research group and obtain funding?	Yes	No
3. Will taking this position enhance my ability to develop productive collaborations and do good work?	Yes	No
4. Is the general atmosphere stimulating?	Yes	No
5. Is the general atmosphere optimistic?	Yes	No
6. Is the general atmosphere competitive?	Yes	No
7. Is the general atmosphere enthusiastic?	Yes	No
8. Is the general atmosphere apathetic?	No	Yes
9. Is the general atmosphere pessimistic?	No	Yes
10. Is the general atmosphere paranoid?	No	Yes
11. Is the general atmosphere crazy?	No	Yes
12. Is the general atmosphere desperate?	No	Yes
13. Is grant support highly valued?	Yes	No
14. Are publications highly valued?	Yes	No
15. Is productivity highly valued?	Yes	No
16. Is teaching highly valued?	Yes	No
17. Do hardworking and productive faculty/staff receive proper recognition and reward?	Yes	No
18. Is competence highly valued?	Yes	No
19. Is reliability highly valued?	Yes	No
20. Is sense of humor highly valued?	Yes	No

21. Is erudition highly valued?   Yes   No

22. Is clique membership highly valued?   No   Yes

23. Is cunning highly valued?   No   Yes

24. Is arrogance highly valued?   No   Yes

25. Is fame highly valued?   Yes   No

26. Is competition mainly constructive?   Yes   No

27. Is competition mainly fair?   Yes   No

28. Is competition mainly helpful?   Yes   No

29. Is competition mainly vigorous?   Yes   No

30. Is competition mainly backbiting?   No   Yes

31. Is competition mainly discouraging?   No   Yes

32. Is competition mainly destructive?   No   Yes

33. Is competition mainly insidious?   No   Yes

34. Is competition mainly exciting?   Yes   No

35. Is competition mainly tiring?   No   Yes

36. Is the physical plant adequate?   Yes   No

37. Are faculty/staff free of minor irritations?   Yes   No

38. Have faculty/staff been victims of major crimes?   Yes   No

39. Is intellectual stimulation adequate?   Yes   No

40. National rank of institution   top 25%   bottom 25%

41. Is general faculty/staff welfare given adequate consideration?   Yes   No

42. Is minority faculty/staff welfare given adequate consideration?   Yes   No

43. Are faculty/staff involved in administration?   Yes   No

44. Are the Dean/President respected for their academic accomplishments?   Yes   No

45. Is it generally known that faculty/staff have filed lawsuits against the administration?   No   Yes

46. Is overall administrative competence at least adequate?   Yes   No

47. Are disputes usually resolved amicably?   Yes   No

48. Are administrative decisions rational?   Yes   No

49. Is the administration "top-heavy?"   No   Yes

50. Is faculty/staff research adequately supported?   Yes   No

51. Are written agreements frequently broken?   No   Yes

52. Are oral agreements frequently broken?   No   Yes

53. Do faculty/staff from within the institution ever interact socially?   Yes   No

54. Score one positive point for each advantage mentioned and one negative point for each disadvantage. Are there advantages of locating at this institution that outweigh the disadvantages?

55. Would you leave to take a position of equal rank and salary elsewhere?   No   Yes

FAMILY LIFE QUESTIONS

                                                   Positive   Negative

F1. Will locating at this institution make raising a family relatively easy?   Yes   No

F2. Have faculty/staff been victims of major crimes?   No   Yes

F3. Have faculty/staff been victims of minor crimes?   No   Yes

F4. Is commuting time from home to work 30 min. or less?   Yes   No

F5. Does commuting significantly interfere with work?

No   Yes

F6. Is above average child care available and affordable?

Yes   No

F7. Are schools above average?

Yes   No

F8. Is above average housing available and affordable?

Yes   No

PRODUCTIVITY QUESTIONS

Responses to the survey questions most highly correlated with production of grants and publications were used to calculate a Productivity P/N score. The questions are repeated here in order to make evaluation of this subset more convenient. Numbers used to identify these questions in the overall academic atmosphere section are in parentheses.

Positive   Negative

P1(1). Will locating at this institution help me move to a better position in the future?

Yes   No

P2(2). Will taking this position enhance my ability to build a research group and obtain funding?

Yes   No

P3(3). Will taking this position enhance my ability to develop productive collaborations and do good work?

Yes   No

P5(4). Is the general atmosphere stimulating?

Yes   No

P6(6). Is the general atmosphere competitive?

Yes   No

P7(25). Is fame highly valued?

Yes   No

P8(29). Is competition mainly vigorous?

Yes   No

P9(39). Is intellectual stimulation adequate?

Yes   No

P11(42). National rank of institution   top 25%   bottom 25%

P/N SCORE CALCULATIONS

Academic Atmosphere

$$\frac{\text{Total Positive}}{\text{Total Negative}} = \underline{\hspace{2cm}} = \boxed{}$$

Family Life

$$\frac{\text{Total Positive}}{\text{Total Negative}} = \underline{\hspace{2cm}} = \boxed{}$$

Productivity

$$\frac{\text{Total Positive}}{\text{Total Negative}} = \underline{\hspace{2cm}} = \boxed{}$$

Appendix II
P/N Ratios for Selected Institutions

The following list of positive/negative response ratios (P/N ratios) is based entirely on faculty and professional staff answers to the questions in Chapter 6. The results indicate that P/N ratios for general institutional atmosphere can be expected to vary over at least a 100-fold range. This is so even though the group of institutions surveyed share many other characteristics. For example, they are mostly medical schools, major universities, and pharmaceutical companies where teaching, research, and professional activities are all biomedically oriented. Applying the evaluation method to other apparently homogeneous groups of faculty or staff, e.g., all faculty at small liberal arts colleges, should also yield a wide range of P/N ratios. However, the boundaries defining highly positive and highly negative academic environments might be somewhat different when the method is applied to different schools. In this case, simply ranking the schools would suffice until more data is available.

Tests of statistical significance using the total number of positive and negative responses for each school indicate that two-fold differences in P/N ratio are generally significant at the 95% level. Smaller differences may become significant as the sample size increases. In this survey, 3 to 12 responses per question were obtained. The effects of sample size were considered in defining the boundaries for highly positive and highly negative P/N ratio. The P/N ratios obtained by this sampling method appear to reflect the overall institutional atmosphere quite accurately because resurveying the entire faculty of one institution, Pennsylvania State University College of Medicine, yielded data that was not significantly different from that obtained by the much smaller original sample.

P/N Ratios and Key to Institution Codes

Institution		Academic Atmosphere		Personal Life and Family		Productivity	
Code	School Name	P/N	Survey Rank	P/N	Survey Rank	P/N	Survey Rank
	U.S. Universities						
	Alabama						
182	Auburn University	1.80	162	41.00	17	0.59	112
1	University of Alabama at Birmingham	12.22	7	23.25	53	2.02	28
2	University of South Alabama School of Medicine	1.39	181	3.43	165	0.40	138
	Arizona						
183	Arizona State University	1.82	161	19.00	64	1.00	79
3	University of Arizona	1.89	155	10.67	104	0.44	129
	Arkansas						
4	University of Arkansas for Medical Sciences	1.71	166	8.00	121	0.25	174
	California						
184	California Institute of Technology	10.90	10	11.00	100	2.71	18
10	Loma Linda University School of Medicine	4.54	60	17.00	71	0.41	137
185	San Diego State University	2.38	123	8.33	117	0.55	116
127	San Francisco VA Medical Center	2.98	95	28.00	42	7.00	3
125	Scripps Clinic and Research Foundation	1.89	155	6.00	145	3.82	7
12	Stanford University	4.72	54	12.50	93	4.20	6
124	Univ. of California at Berkeley	2.98	95	2.67	175	2.85	14
5	Univ. of California at Davis	1.94	149	7.75	124	0.37	147
6	Univ. of California at Irvine	6.72	24	9.20	109	1.00	78
7	Univ. of California at Los Angeles	14.11	5	1.00	204	1.94	31
126	Univ. of California at Riverside	2.31	126	8.33	117	0.34	153
8	Univ. of California at San Diego	4.33	65	5.67	148	1.09	66

9	Univ. of California at San Francisco	5.03	48	1.24	201	2.58	19
128	Univ. of California at Santa Barbara	3.50	80	4.67	154	1.18	61
11	University of Southern California	4.14	66	1.43	196	0.42	134
	Colorado						
193	Colorado State University	2.42	120	45.00	14	0.74	92
129	University of Colorado at Boulder	3.77	75	28.00	42	1.24	56
13	University of Colorado School of Medicine	2.93	97	14.00	85	0.87	85
130	Natl. Jewish Center for Immunol. & Resp. Med.	4.12	67	20.50	60	1.96	30
	Connecticut						
131	University of Connecticut at Storrs	5.68	37	46.00	10	1.00	75
14	University of Connecticut School of Medicine	2.09	138	13.50	88	0.31	163
15	Yale University	1.97	145	1.27	200	1.88	34
	Delaware						
190	University of Delaware at Newark	2.08	139	14.00	85	0.54	119
	District of Columbia						
16	George Washington University School of Medicine	1.95	147	1.50	194	0.37	145
17	Georgetown University School of Medicine	0.45	207	2.67	175	0.28	171
18	Howard University	0.59	205	1.78	187	0.22	183
192	Walter Reed Army Institute of Research	2.17	131	17.00	71	0.88	84
	Florida						
194	Florida State University	3.44	83	12.00	95	0.59	113
19	University of Florida at Gainesville	4.00	71	8.25	119	0.71	100
20	University of Miami	1.64	170	1.79	186	0.39	142
21	University of South Florida School of Medicine	1.57	173	4.17	157	0.15	201
	Georgia						
22	Emory University	2.15	133	6.40	136	0.30	164
195	Georgia Institute of Technology	1.97	145	11.00	100	0.63	106
23	Medical College of Georgia	1.32	183	4.86	152	0.33	157
24	Mercer University School of Medicine	1.84	160	5.00	149	0.18	196

P/N Ratios and Key to Institution Codes (*Continued*)

Institution		Academic Atmosphere		Personal Life and Family		Productivity	
Code	School Name	P/N	Survey Rank	P/N	Survey Rank	P/N	Survey Rank
25	Morehouse School of Medicine	1.62	171	2.60	178	0.22	185
132	University of Georgia at Athens	5.81	33	46.00	10	1.55	43
Hawaii							
133	University of Hawaii at Manoa	1.44	178	6.40	136	0.29	168
Illinois							
27	Chicago Medical School	4.61	56	16.00	77	0.25	175
29	Loyola University School of Medicine	0.91	199	3.29	169	0.20	193
134	Northwestern University at Evanston	1.91	153	15.00	82	1.13	64
30	Northwestern University School of Medicine	1.48	175	2.00	183	0.52	123
31	Rush Medical College	1.37	182	1.74	189	0.32	159
32	Southern Illinois University School of Medicine	0.83	200	38.00	24	0.19	194
26	University of Chicago	2.75	106	1.29	199	1.08	67
135	University of Illinois at Urbana-Champaign	14.87	4	17.00	70	3.36	11
28	University of Illinois at Chicago	1.02	196	0.83	208	0.21	190
Indiana							
136	Indiana University at Bloomington	7.14	21	14.50	84	1.47	50
33	Indiana University School of Medicine	6.33	29	8.50	115	0.33	158
137	Purdue University	3.80	74	17.00	70	1.05	70
198	University of Notre Dame	2.02	143	23.00	54	0.40	139
Iowa							
138	Iowa State University	3.66	77	20.00	62	1.42	51
34	University of Iowa	40.00	1	47.00	9	0.88	83

	Kansas						
140	Kansas State University	2.33	125	40.00	21	0.14	205
139	University of Kansas at Lawrence	2.19	129	41.00	17	0.60	111
35	University of Kansas School of Medicine	1.11	190	7.50	126	0.23	179
	Kentucky						
36	University of Kentucky School of Medicine	10.41	11	58.00	3	0.49	127
37	University of Louisville School of Medicine	3.71	76	11.50	98	0.22	184
	Louisiana						
141	Louisiana State University at Baton Rouge	1.45	176	4.00	159	0.50	124
38	Louisiana State University at New Orleans	0.93	198	1.14	202	0.32	161
39	Louisiana State University at Shreveport	1.43	179	3.42	166	0.18	198
40	Tulane University School of Medicine	2.12	135	1.60	191	0.22	182
	Maryland						
199	FDA Center for Biologics	2.56	113	21.00	58	1.00	76
41	Johns Hopkins University School of Medicine	5.32	41	1.55	193	2.81	16
143	National Institutes of Health	2.89	100	43.00	16	1.87	35
145	NCI Frederick Cancer Research Center	9.70	14	16.00	77	2.71	17
144	University of Maryland at College Park	1.27	186	3.83	162	0.50	125
42	University of Maryland School of Medicine	2.39	122	2.33	180	0.45	128
43	Uniformed Services University School of Medicine	5.78	36	11.33	99	0.35	152
142	USDA at Beltsville	2.10	137	11.00	100	0.55	117
	Massachusetts						
149	Brandeis University	1.92	152	27.00	46	1.22	58
44	Boston University School of Medicine	2.84	102	3.25	170	0.42	136
45	Harvard University School of Medicine	4.52	61	1.78	187	3.68	8
147	Massachusetts Institute of Technology	26.25	2	8.00	121	23.00	1
146	University of Massachusetts at Amherst	4.55	58	29.00	39	2.18	22
46	University of Massachusetts School of Medicine	3.53	79	6.67	135	1.19	60
148	Worcester Foundation for Experimental Biology	3.26	86	40.00	21	2.11	26
47	Tufts University School of Medicine	5.24	43	2.33	180	0.71	99

P/N Ratios and Key to Institution Codes *(Continued)*

	Institution	Academic Atmosphere		Personal Life and Family		Productivity	
Code	School Name	P/N	Survey Rank	P/N	Survey Rank	P/N	Survey Rank
	Michigan						
48	Michigan State University	4.42	63	52.00	6	0.42	133
49	University of Michigan	4.55	58	7.13	131	1.58	41
50	Wayne State University	6.32	30	2.86	174	0.61	110
	Minnesota						
51	Mayo Medical School	3.18	89	37.00	25	0.71	98
150	Minneapolis VA Medical Center	1.98	144	28.00	42	0.54	118
52	University of Minnesota at Duluth School of Medicine	6.56	26	48.00	8	0.27	172
	University of Minnesota at Minneapolis School of						
53	Medicine	6.34	28	6.90	133	1.51	46
151	University of Minnesota at St. Paul	3.23	88	15.00	82	0.94	81
	Mississippi						
54	University of Mississippi School of Medicine	4.70	55	52.00	6	0.44	131
	Missouri						
55	University of Missouri at Columbia	2.48	116	21.00	58	0.36	149
56	University of Missouri at Kansas City School of Medicine	2.11	136	11.67	97	0.19	195
57	St. Louis University School of Medicine	3.02	93	7.60	125	0.39	143
58	Washington University (St. Louis) School of Medicine	7.77	20	6.17	141	2.36	21
	Nebraska						
59	Creighton University School of Medicine	0.54	206	6.40	136	0.13	208
152	University of Nebraska at Lincoln	5.81	33	18.00	68	1.17	63
60	University of Nebraska School of Medicine	4.05	70	12.33	94	0.76	91

	Nevada						
61	University of Nevada School of Medicine	2.76	105	3.10	172	0.38	144
	New Hampshire						
62	Dartmouth Medical School	1.95	147	33.00	32	0.30	165
	New Jersey						
206	Princeton University	6.89	23	17.00	70	10.33	2
208	Rutgers University	2.42	120	11.00	100	1.07	68
63	UMDNJ at Newark	0.82	201	1.35	197	0.18	197
64	UMDNJ at Piscataway (R. W. Johnson Medical School)	3.48	82	6.13	144	0.34	154
	New Mexico						
65	University of New Mexico School of Medicine	7.00	22	5.00	149	0.49	126
	New York						
66	Albany Medical College	1.08	192	22.67	55	0.25	176
67	Albert Einstein University School of Medicine	9.58	15	4.64	155	1.91	33
209	Brookhaven National Laboratory	5.83	32	19.00	64	1.60	40
68	Columbia University College of Physicians & Surgeons	1.71	166	1.00	204	2.13	24
155	Cornell University at Ithaca	10.33	12	20.00	62	5.60	4
69	Cornell University School of Medicine	2.20	128	0.93	206	1.03	74
157	Memorial-Sloan Kettering	5.42	39	7.50	126	3.25	12
70	Mount Sinai School of Medicine	5.00	49	1.11	203	1.27	54
71	New York Medical College	1.31	184	8.80	113	0.28	170
72	New York University School of Medicine	2.15	132	1.93	184	0.62	107
211	NYS Institute for Basic Research	2.18	130	12.00	95	1.00	77
156	Rockefeller University	5.97	31	1.91	185	4.67	5
154	Roswell Park Cancer Institute	1.42	180	22.00	56	0.86	87
74	SUNY at Brooklyn School of Medicine	0.36	208	0.86	207	0.13	206
75	SUNY at Buffalo	1.45	176	8.83	112	0.29	167
76	SUNY at Stony Brook	2.23	127	9.00	111	0.22	186
77	SUNY at Syracuse School of Medicine	2.07	141	10.20	107	0.21	189
73	University of Rochester School of Medicine	8.55	16	15.20	81	1.10	65

P/N Ratios and Key to Institution Codes *(Continued)*

Institution		Academic Atmosphere		Personal Life and Family		Productivity	
Code	School Name	P/N	Survey Rank	P/N	Survey Rank	P/N	Survey Rank
153	Wadsworth Center NYS Health Department	5.08	45	25.00	51	1.79	36
	North Carolina						
78	Bowman Gray School of Medicine	2.08	139	7.20	129	0.43	132
79	Duke University School of Medicine	2.73	107	6.40	136	1.52	45
80	East Carolina University School of Medicine	6.62	25	13.33	89	0.23	177
158	North Carolina State University	3.50	80	9.25	108	2.05	27
159	Natl. Inst. Environmental Health Sciences	2.90	98	27.00	46	1.50	48
81	University of North Carolina School of Medicine	8.15	19	33.50	31	1.04	71
	North Dakota						
82	University of North Dakota School of Medicine	1.94	149	32.00	32	0.17	200
	Ohio						
83	Case Western Reserve University	3.13	90	5.00	149	0.66	104
217	Cincinnati Children's Hospital	4.50	62	34.00	29	1.23	57
214	Cleveland Clinic Foundation	3.96	72	13.00	90	2.00	29
84	University of Cincinnati	2.66	110	8.00	121	0.74	93
85	Medical College of Ohio	0.96	197	28.50	40	0.10	209
86	Northeastern Ohio University School of Medicine	2.53	115	28.50	40	0.14	204
87	Ohio State University	3.38	85	6.40	136	0.81	89
88	Wright State University	1.06	194	32.00	32	0.13	207
		(1.65)	(170)	(10.00)	(107)	(0.23)	(180)
	Oklahoma						
160	Oklahoma Medical Research Foundation	2.81	103	2.40	179	1.30	53
89	University of Oklahoma School of Medicine	1.87	158	6.14	143	0.21	187

90	Oral Roberts University School of Medicine	2.67	109	24.00	52	0.23	180

Oregon

91	Oregon Health Science University	8.24	18	41.00	17	0.42	135
161	Oregon State University	3.11	91	53.00	4	0.59	114
218	University of Oregon	2.71	108	34.00	29	1.71	37

Puerto Rico

| 213 | University of Puerto Rico School of Medicine | 0.54 | 203 | 1.60 | 191 | 0.07 | 210 |

Pennsylvania

212	Carnegie Mellon University	-11.07	9	26.00	48	3.56	9
92	Hahnemann University School of Medicine	1.14	189	3.54	164	0.23	178
93	Thomas Jefferson University School of Medicine	5.11	44	3.92	161	0.53	122
94	Medical College of Pennsylvania	2.90	98	3.25	170	0.61	109
95	Pennsylvania State University School of Medicine	1.90	154	45.00	14	0.21	188
		(2.84)	(102)	(38.00)	(24)	(0.79)	(90)
162	Pennsylvania State University at University Park	5.58	38	33.00	32	1.67	38
96	University of Pennsylvania	3.00	94	1.67	190	2.13	23
97	University of Pittsburgh	1.07	193	7.20	129	0.37	146
98	Temple University School of Medicine	4.77	52	2.15	182	0.67	103

Rhode Island

| 99 | Brown University | 12.50 | 6 | 3.00 | 173 | 0.71 | 97 |

South Carolina

| 100 | Medical University of South Carolina | 0.75 | 202 | 6.17 | 141 | 0.20 | 193 |
| 101 | University of South Carolina School of Medicine | 4.56 | 57 | 5.89 | 147 | 0.34 | 155 |

South Dakota

| 102 | University of South Dakota School of Medicine | 2.58 | 112 | 67.00 | 2 | 0.17 | 199 |

Tennessee

103	East Tennessee State University	1.67	169	9.14	110	0.06	211
104	Meharry Medical College	0.67	204	3.33	168	0.15	202
221	Oak Ridge National Laboratory	2.54	114	13.00	90	0.62	108
220	St. Jude Children's Research Hospital	5.81	33	16.00	77	1.21	59

P/N Ratios and Key to Institution Codes (Continued)

Institution		Academic Atmosphere		Personal Life and Family		Productivity	
Code	School Name	P/N	Survey Rank	P/N	Survey Rank	P/N	Survey Rank
163	University of Tennessee at Knoxville	1.09	191	6.00	145	0.28	169
105	University of Tennessee School of Medicine	1.86	159	4.00	159	0.72	96
106	Vanderbilt University School of Medicine	4.11	69	3.78	163	3.38	10
	Texas						
107	Baylor College of Medicine	1.75	163	3.38	167	0.66	105
222	Rice University	16.67	3	7.00	132	2.13	25
108	Texas A & M University	1.94	149	40.00	21	0.35	150
165	University of North Texas Health Sciences Center	2.13	134	17.00	70	0.40	140
109	Texas Technical University	1.71	166	18.50	67	0.32	160
110	University of Texas Southwestern Medical Center	9.95	13	4.86	152	2.94	13
164	University of Texas at Austin	2.37	124	41.00	17	1.03	73
111	University of Texas at Galveston School of Medicine	1.73	165	1.44	195	0.33	156
112	University of Texas at Houston School of Medicine	1.03	195	1.31	198	0.40	141
166	University of Texas-M. D. Anderson Cancer Center	2.44	118	4.50	156	1.17	62
113	University of Texas at San Antonio School of Medicine	2.66	110	8.25	119	0.27	173
	Utah						
114	University of Utah	3.63	78	8.75	114	0.53	120
223	Utah State University	1.26	187	37.00	25	0.29	166
	Vermont						
115	University of Vermont School of Medicine	4.81	51	13.00	90	0.87	86
	Virginia						
116	Eastern Virginia Medical School	4.84	50	7.50	126	0.35	151

117	Medical College of Virginia	1.16	188	2.62	177	0.58	115
118	University of Virginia School of Medicine	11.68	8	16.67	76	0.95	80
167	VPI and State University	3.26	86	26.00	48	0.68	101

Washington

224	Fred Hutchinson Cancer Research Center	8.52	17	28.00	42	2.85	15
119	University of Washington	5.04	47	4.17	157	1.50	47
168	Washington State University	4.37	64	30.00	38	1.06	69

West Virginia

120	Marshall University School of Medicine	5.08	45	31.00	36	0.23	181
121	West Virginia University of School of Medicine	1.54	174	69.00	1	0.20	191

Wisconsin

122	Medical College of Wisconsin	2.88	101	26.00	48	0.68	102
169	Milwaukee VA Medical Center	2.48	116	17.50	69	1.25	55
123	University of Wisconsin	5.42	39	6.86	134	1.63	39

U.S. Liberal Arts Colleges

	U.S. Liberal Arts Colleges	3.80	74	17.13	70	0.72	95

Canadian Medical Schools

174	Dalhousie University School of Medicine	1.60	172	15.33	80	0.36	148
179	McGill University School of Medicine	6.45	27	10.67	104	1.40	52
176	McMaster University School of Medicine	4.12	67	46.00	10	1.92	32
177	Queen's University School of Medicine	5.28	42	35.00	28	1.04	72
171	University of Alberta School of Medicine	4.73	53	53.00	4	1.48	49
172	University of British Columbia School of Medicine	2.06	142	8.50	115	0.90	82
170	University of Calgary School of Medicine	3.95	73	36.00	27	0.78	90
175	University of Guelph School of Medicine	2.43	119	21.50	57	0.83	88
173	University of Manitoba School of Medicine	2.78	104	19.00	64	0.53	121

P/N Ratios and Key to Institution Codes (*Continued*)

Institution		Academic Atmosphere		Personal Life and Family		Productivity	
Code	School Name	P/N	Survey Rank	P/N	Survey Rank	P/N	Survey Rank
178	University of Ottawa School of Medicine	1.89	155	13.67	87	0.32	162
180	University of Saskatchewan School of Medicine	1.28	185	20.50	60	0.15	203
181	University of Toronto School of Medicine	3.02	92	10.50	106	2.45	20
225	University of Western Ontario	1.75	163	31.00	36	0.73	94
	Pharmaceutical and Industrial Research Institutes						
202	Parke-Davis Pharmaceutical Res.	3.40	84	45.33	13	1.54	44

Medical Schools are included with the parent university unless otherwise indicated.

Appendix III

Publication Rates for Selected Institutions

Publication counts for the 1993 calendar year were obtained from the National Library of Medicine (MEDLINE) database using CDPLUS: the "institution" field was searched for both institution name and the appropriate ZIP code(s), the data set was limited to conventional articles, and the first 25 citations were examined manually to verify the validity of the search criteria. This method yielded more complete data than manually counting papers in 30 major journals (the technique used for the first edition) and can be more easily verified and updated. The number of faculty qualified to publish in professional journals at each institution was taken as the total number of members listed in the 1993/1994 Directory of the Federation of Societies for Experimental Biology (FASEB). This estimate was assumed to be proportional to the true value, which is obviously larger.

Institutional and Per Capita Publication Rates

Code	FASEB Members	Total Pubs		Pubs/FASEB Member		Most Highly Cited	
		Number	Rank (N = 223)	Pubs/F	Rank (N = 210)	Number	Rank (N = 223)
1	238	760	34	3.19	116	3	51
2	75	145	156	1.93	172	0	NA
3	163	625	45	3.83	81	3	51
4	80	254	116	3.17	118	4	45
5	222	975	23	4.39	54	5	39
6	101	517	64	5.12	28	0	NA
7	273	1359	8	4.98	35	18	13
8	248	1030	20	4.15	63	6	31
9	272	1910	3	7.02	9	28	5
10	45	205	128	4.56	48	0	NA
11	131	889	26	6.79	11	7	26
12	146	1186	14	8.12	7	20	11
13	152	618	47	4.07	69	19	12
14	104	359	87	3.45	102	1	73
15	249	1329	9	5.34	24	15	16
16	48	229	120	4.77	40	2	61
17	98	429	77	4.38	55	4	45
18	58	83	185	1.43	190	0	NA
19	184	924	24	5.02	31	4	45
20	122	605	52	4.96	37	1	73
21	82	382	82	4.66	46	0	NA
22	166	681	38	4.10	66	3	51
23	99	330	92	3.33	105	1	73
24	19	36	216	1.89	177	0	NA

25	26	27	220	1.04	204	0	NA
26	163	774	33	4.75	41	12	18
27	53	90	184	1.70	182	1	73
28	175	619	46	3.54	99	6	31
29	71	362	85	5.10	29	1	73
30	128	524	63	4.09	67	4	45
31	75	319	96	4.25	59	0	NA
32	29	60	195	2.07	167	0	NA
33	122	640	44	5.25	26	2	61
34	204	1036	19	5.08	30	2	61
35	102	317	97	3.11	124	3	51
36	167	451	73	2.70	142	1	73
37	93	338	90	3.63	92	0	NA
38	104	284	105	2.73	138	1	73
39	71	289	104	4.07	68	0	NA
40	92	153	153	1.66	184	4	45
41	332	1828	4	5.51	20	26	6
42	157	503	66	3.20	115	3	51
43	71	196	130	2.76	137	1	73
44	197	536	61	2.72	139	8	25
45	483	2944	2	6.10	15	92	1
46	90	361	86	4.01	72	0	NA
47	142	561	59	3.95	76	9	23
48	135	443	76	3.28	112	1	73
49	183	1609	6	8.79	2	13	17
50	166	887	27	5.34	23	9	23
51	142	1212	12	8.54	4	6	31
52	21	28	219	1.33	193	5	39
53	237	1090	17	4.60	47	11	19
54	74	191	133	2.58	147	0	NA

Institutional and Per Capita Publication Rates *(Continued)*

Code	FASEB Members	Total Pubs		Pubs/FASEB Member		Most Highly Cited	
		Number	Rank (N = 223)	Pubs/F	Rank (N = 210)	Number	Rank (N = 223)
55	121	422	79	3.49	101	1	73
56	44	157	151	3.57	96	0	NA
57	87	290	102	3.33	103	0	NA
58	239	1291	10	5.40	22	18	13
59	47	185	139	3.94	78	0	NA
60	82	320	95	3.90	79	0	NA
61	49	93	180	1.90	176	0	NA
62	82	255	115	3.11	123	0	NA
63	104	292	101	2.81	134	0	NA
64	77	295	100	3.83	82	0	NA
65	73	266	110	3.64	91	0	NA
66	58	187	136	3.22	113	1	73
67	155	659	40	4.25	60	10	22
68	189	986	22	5.22	27	24	8
69	130	586	55	4.51	49	18	13
70	120	704	37	5.87	18	11	19
71	79	188	134	2.38	152	1	73
72	144	584	56	4.06	70	23	9
73	133	652	43	4.90	38	7	26
74	66	272	108	4.12	65	2	61
75	153	460	71	3.01	128	1	73
76	127	451	74	3.55	97	1	73
77	84	151	154	1.80	181	0	NA
78	106	400	80	3.77	84	0	NA

79	213	1200	13	5.63	19	5	39
80	65	195	131	3.00	129	0	NA
81	224	1054	18	4.71	43	2	61
82	49	78	187	1.59	188	0	NA
83	154	836	29	5.43	21	7	26
84	144	611	50	4.24	61	5	39
85	69	159	150	2.30	156	0	NA
86	27	59	198	2.19	163	0	NA
87	198	923	25	4.66	45	1	73
88	53	105	176	1.98	168	0	NA
89	77	304	99	3.95	77	0	NA
90	1	1	223	1.00	206	0	NA
91	98	598	53	6.10	14	2	61
92	57	188	135	3.30	107	2	61
93	134	790	32	5.90	17	0	NA
94	52	138	162	2.65	144	0	NA
95	88	290	103	3.30	108	0	NA
96	255	1662	5	6.52	12	22	10
97	179	1254	11	7.01	10	6	31
98	122	330	91	2.70	141	1	73
99	65	324	93	4.98	34	2	61
100	103	389	81	3.78	83	0	NA
101	63	182	143	2.89	131	0	NA
102	27	24	221	0.89	209	0	NA
103	39	82	186	2.10	166	0	NA
104	32	34	217	1.06	201	0	NA
105	114	444	75	3.89	80	2	61
106	196	722	36	3.68	89	5	39
107	203	1010	21	4.98	36	6	31

Institutional and Per Capita Publication Rates *(Continued)*

Code	FASEB Members	Total Pubs		Pubs/FASEB Member		Most Highly Cited	
		Number	Rank (N = 223)	Pubs/F	Rank (N = 210)	Number	Rank (N = 223)
108	108	342	88	3.17	120	0	NA
109	75	147	155	1.96	171	0	NA
110	234	823	30	3.52	100	3	51
111	168	512	65	3.05	127	0	NA
112	129	611	49	4.74	42	3	51
113	154	658	41	4.27	58	0	NA
114	138	613	48	4.44	52	1	73
115	91	273	107	3.00	130	1	73
116	42	133	165	3.17	121	0	NA
117	146	588	54	4.03	71	2	61
118	178	657	42	3.69	87	3	51
119	248	1486	7	5.99	16	25	7
120	21	22	222	1.05	202	0	NA
121	93	221	121	2.38	153	0	NA
122	100	531	62	5.31	25	0	NA
123	264	1183	15	4.48	50	32	4
124	150	667	39	4.45	51	6	31
125	145	456	72	3.14	122	3	51
126	39	144	157	3.69	86	0	NA
127	99	106	175	1.07	200	0	NA
128	35	92	181	2.63	145	0	NA
129	70	159	149	2.27	158	0	NA
130	34	144	158	4.24	62	0	NA

131	49	136	164	2.78	136	0	NA
132	89	320	94	3.60	95	1	73
133	53	192	132	3.62	93	0	NA
134	43	185	140	4.30	57	0	NA
135	130	483	68	3.72	85	0	NA
136	42	166	147	3.95	75	1	73
137	114	375	84	3.29	110	0	NA
138	74	186	137	2.51	149	0	NA
139	32	91	183	2.84	132	0	NA
140	48	177	144	3.69	88	0	NA
141	52	123	171	2.37	154	0	NA
142	55	140	160	2.55	148	76	2
143	715	2958	1	4.14	64	0	NA
144	35	169	146	4.83	39	0	NA
145	82	271	109	3.30	106	0	NA
146	37	131	166	3.54	98	7	26
147	76	379	83	4.99	33	0	NA
148	38	45	209	1.18	196	2	61
149	53	102	179	1.92	173	0	NA
150	31	102	178	3.29	109	0	NA
151	66	212	126	3.21	114	0	NA
152	40	112	173	2.80	135	0	NA
153	43	72	190	1.67	183	4	45
154	67	220	122	3.28	111	0	NA
155	127	464	70	3.65	90	37	3
156	99	305	98	3.08	125	6	31
157	67	571	58	8.52	5	0	NA
158	65	207	127	3.18	117	0	NA
159	48	241	118	5.02	32	0	NA

Institutional and Per Capita Publication Rates *(Continued)*

Code	FASEB Members	Total Pubs		Pubs/FASEB Member		Most Highly Cited	
		Number	Rank (N = 223)	Pubs/F	Rank (N = 210)	Number	Rank (N = 223)
160	43	59	196	1.37	192	0	NA
161	64	137	163	2.14	164	0	NA
162	95	220	123	2.32	155	0	NA
163	68	155	152	2.28	157	0	NA
164	83	263	111	3.17	119	1	73
165	44	52	201	1.18	197	0	NA
166	86	748	35	8.70	3	0	NA
167	47	186	138	3.96	74	1	73
168	85	204	129	2.40	151	0	NA
169	23	38	213	1.65	185	0	NA
170	70	474	69	NA	NA	0	NA
171	89	609	51	NA	NA	0	NA
172	98	842	28	NA	NA	1	73
173	86	429	78	NA	NA	0	NA
174	38	273	106	NA	NA	0	NA
175	51	234	119	NA	NA	0	NA
176	66	541	60	NA	NA	1	73
177	41	250	117	NA	NA	0	NA
178	41	259	114	NA	NA	0	NA
179	91	806	31	NA	NA	5	39
180	37	261	112	NA	NA	0	NA
181	109	1179	16	NA	NA	11	19
182	48	128	169	2.67	143	0	NA

183	28	123	170	4.39	53	0	NA
184	37	160	148	4.32	56	7	26
185	58	73	189	1.26	194	0	NA
186	33	63	193	1.91	175	0	NA
188	68	109	174	1.60	187	0	NA
189	57	50	203	0.88	210	0	NA
190	28	76	188	2.71	140	0	NA
191	26	29	218	1.12	199	1	73
192	41	261	113	6.37	13	0	NA
193	73	183	142	2.51	150	0	NA
194	37	61	194	1.65	186	0	NA
195	10	47	205	4.70	44	0	NA
196	66	130	167	1.97	170	0	NA
197	77	173	145	2.25	160	0	NA
198	22	41	211	1.86	178	0	NA
199	40	45	208	1.13	198	0	NA
200	21	46	207	2.19	162	0	NA
201	83	184	141	2.22	161	0	NA
202	67	92	182	1.37	191	0	NA
203	52	50	202	0.96	207	0	NA
204	36	37	214	1.03	205	0	NA
205	44	46	206	1.05	203	0	51
206	36	130	168	3.61	94	3	NA
207	92	143	159	1.55	189	0	73
208	111	341	89	3.07	126	1	31
209	24	54	200	2.25	159	6	NA
211	23	44	210	1.91	174	0	NA
212	32	59	197	1.84	180	0	NA
213	32	59	199	1.84	179	0	NA
214	69	584	57	8.46	6	1	73

Institutional and Per Capita Publication Rates *(Continued)*

Code	FASEB Members	Total Pubs		Pubs/FASEB Member		Most Highly Cited	
		Number	Rank (N = 223)	Pubs/F	Rank (N = 210)	Number	Rank (N = 223)
215	34	41	212	1.21	195	0	NA
216	41	37	215	0.90	208	0	NA
217	31	123	172	3.97	73	0	NA
218	36	102	177	2.83	133	0	NA
219	65	138	161	2.12	165	0	NA
220	27	214	125	7.93	8	0	NA
221	34	67	192	1.97	169	2	61
222	18	47	204	2.61	146	0	NA
223	21	70	191	3.33	104	0	NA
224	19	218	124	11.47	1	0	NA
225	71	499	67	NA	NA	0	NA
Liberal Arts	103	172	NA	1.67	NA	1	NA
Totals	21,787	92,050				805	

Appendix IV

NIH Grant Support for Selected Institutions

Grant support data for fiscal year 1993 were obtained from "National Institutes of Health Research Grants and Contracts" on GOPHER. Data were combined for geographic and administrative subdivisions of certain institutions if, in the author's judgment, they comprised a single academic entity.

Number of faculty qualified to apply for NIH grants at each institution was taken as the total number of members listed in the 1993/1994 Directory of the Federation of Societies for Experimental Biology (FASEB). This estimate was assumed to be proportional to the true value, which is obviously larger. Thus, NIH grant support per capita is overestimated, but proportional to the true value and useful for interinstitutional comparisons. The total number of FASEB members in this table is slightly smaller than that in Appendix III because institutions for which NIH grant support is not a valid measure of productivity are not included.

Institutional and Per Capita Grant Support

Code	FASEB Members	No. of Grants	G/F	Rank ($N = 181$)	Funding M$	M$/F	Rank ($N = 181$)
1	238	312	1.31	58	69.8	0.29	57
2	75	40	0.53	152	7.2	0.10	146
3	163	198	1.21	70	48.4	0.30	54
4	80	68	0.85	111	11	0.14	125
5	222	203	0.91	104	56	0.25	72
6	101	154	1.52	41	36	0.36	42
7	273	485	1.78	27	130	0.48	19
8	248	404	1.63	33	117	0.47	20
9	272	595	2.19	8	171	0.63	11
10	45	13	0.29	176	2.5	0.06	169
11	131	238	1.82	26	68	0.52	15
12	146	398	2.73	3	120	0.82	2
13	152	245	1.61	34	65	0.43	28
14	104	102	0.98	95	23	0.22	86
15	249	519	2.08	12	141	0.57	13
16	48	51	1.06	85	12	0.25	73
17	98	126	1.29	63	29	0.30	55
18	58	23	0.40	169	7.9	0.14	126
19	184	255	1.39	47	39	0.21	91
20	122	165	1.35	52	49	0.40	31
21	82	51	0.62	141	6.6	0.08	161
22	166	261	1.57	38	63	0.38	38
23	99	66	0.67	136	9.1	0.09	151
24	19	6	0.32	174	0.49	0.03	177
25	26	15	0.58	146	6.6	0.25	70
26	163	258	1.58	37	67	0.41	30
27	53	29	0.55	150	4.9	0.09	150
28	175	173	0.99	94	34	0.19	97
29	71	43	0.61	143	6.1	0.09	153
30	128	124	0.97	97	31	0.24	79
31	75	63	0.84	113	13	0.17	106
32	29	13	0.45	161	1.9	0.07	164
33	122	162	1.33	55	39	0.32	50
34	204	270	1.32	56	66	0.32	48
35	102	86	0.84	112	15	0.15	122
36	167	156	0.93	102	25	0.15	120
37	93	40	0.43	164	5.3	0.06	168
38	104	66	0.63	140	13	0.13	134
39	71	37	0.52	155	5.9	0.08	158
40	92	60	0.65	138	21	0.23	83
41	332	745	2.24	7	224	0.67	5
42	157	196	1.25	66	39	0.25	74
43	71	NA	NA	NA	NA	NA	NA

Institutional and Per Capita Grant Support *(Continued)*

Code	FASEB Members	No. of Grants	G/F	Rank (N = 181)	Funding M$	M$/F	Rank (N = 181)
44	197	227	1.15	76	57	0.29	58
45	483	1044	2.16	9	312	0.65	9
46	90	129	1.43	43	30	0.33	46
47	142	195	1.37	49	44	0.31	52
48	135	113	0.84	114	20	0.15	121
49	183	555	3.03	2	142	0.78	3
50	166	176	1.06	87	34	0.20	96
51	142	172	1.21	71	44	0.31	51
52	21	3	0.14	181	0.27	0.01	181
53	237	434	1.83	24	92	0.39	35
54	74	37	0.50	158	5.3	0.07	163
55	121	96	0.79	117	12	0.10	143
56	44	23	0.52	154	3.7	0.08	155
57	87	68	0.78	120	12	0.14	124
58	239	417	1.74	28	120	0.50	17
59	47	16	0.34	173	3.7	0.08	162
60	82	76	0.93	103	11	0.13	127
61	49	25	0.51	156	5.2	0.11	141
62	82	114	1.39	46	28	0.34	44
63	104	73	0.70	130	16	0.15	118
64	77	86	1.12	79	17	0.22	87
65	73	78	1.07	84	13	0.18	105
66	58	48	0.83	116	7.4	0.13	133
67	155	205	1.32	57	67	0.43	27
68	189	399	2.11	11	122	0.65	10
69	130	170	1.31	59	55	0.42	29
70	120	172	1.43	42	47	0.39	33
71	79	52	0.66	137	13	0.16	113
72	144	239	1.66	29	65	0.45	24
73	133	262	1.97	16	62	0.47	21
74	66	44	0.67	134	14	0.21	90
75	153	133	0.87	108	20	0.13	128
76	127	165	1.30	62	33	0.26	68
77	84	34	0.40	167	7	0.08	156
78	106	142	1.34	53	34	0.32	49
79	213	410	1.92	20	110	0.52	16
80	65	30	0.46	160	3.2	0.05	173
81	224	409	1.83	25	101	0.45	25
82	49	9	0.18	180	1.1	0.02	178
83	154	308	2.00	14	71	0.46	22
84	144	138	0.96	99	38	0.26	65
85	69	42	0.61	142	6.7	0.10	144
86	27	12	0.44	162	1	0.04	174

Institutional and Per Capita Grant Support *(Continued)*

Code	FASEB Members	No. of Grants	G/F	Rank ($N = 181$)	Funding M$	M$/F	Rank ($N = 181$)
87	198	210	1.06	86	37	0.19	99
88	53	23	0.43	163	5.6	0.11	142
89	77	64	0.83	115	10	0.13	130
90	1	NA	NA	NA	NA	NA	NA
91	98	189	1.93	19	35	0.36	41
92	57	44	0.77	122	8	0.14	123
93	134	145	1.08	83	36	0.27	63
94	52	64	1.23	68	15	0.29	59
95	88	103	1.17	73	20	0.23	84
96	255	496	1.95	18	135	0.53	14
97	179	413	2.31	6	117	0.65	8
98	122	95	0.78	121	20	0.16	114
99	65	66	1.02	90	12	0.18	101
100	103	90	0.87	107	16	0.16	117
101	63	44	0.70	132	5.7	0.09	152
102	27	5	0.19	179	0.35	0.01	180
103	39	9	0.23	177	0.81	0.02	179
104	32	24	0.75	124	8.4	0.26	66
105	114	107	0.94	101	19	0.17	112
106	196	269	1.37	50	69	0.35	43
107	203	288	1.42	45	74	0.36	39
108	108	93	0.86	109	14	0.13	132
109	75	22	0.29	175	2.4	0.03	175
110	234	236	1.01	91	61	0.26	67
111	168	152	0.90	105	28	0.17	111
112	129	168	1.30	60	32	0.25	76
113	154	172	1.12	80	43	0.28	60
114	138	225	1.63	32	45	0.33	47
115	91	106	1.16	74	25	0.27	61
116	42	23	0.55	149	3.5	0.08	157
117	146	158	1.08	82	37	0.25	71
118	178	221	1.24	67	48	0.27	62
119	248	532	2.15	10	155	0.63	12
120	21	12	0.57	147	1.1	0.05	171
121	93	37	0.40	168	5.4	0.06	167
122	100	115	1.15	77	24	0.24	80
123	264	435	1.65	31	101	0.38	36
124	150	203	1.35	51	50	0.33	45
125	145	240	1.66	30	72	0.50	18
126	39	47	1.21	72	7.3	0.19	98
127	99	NA	NA	NA	NA	NA	NA
128	35	85	2.43	5	6.3	0.18	104
129	70	91	1.30	61	17	0.24	78

Institutional and Per Capita Grant Support *(Continued)*

Code	FASEB Members	No. of Grants	G/F	' Rank (N = 181)	Funding M$	M$/F	Rank (N = 181)
130	34	54	1.59	36	13	0.38	37
131	49	26	0.53	153	4.2	0.09	154
132	89	70	0.79	119	10	0.11	137
133	53	42	0.79	118	12	0.23	85
134	43	84	1.95	17	17	0.40	32
135	130	130	1.00	92	20	0.15	119
136	42	80	1.90	22	13	0.31	53
137	114	126	1.11	81	21	0.18	102
138	74	40	0.54	151	6.9	0.09	149
139	32	39	1.22	69	6.7	0.21	93
140	48	19	0.40	170	2.6	0.05	170
141	52	19	0.37	172	2.7	0.05	172
142	55	NA	NA	NA	NA	NA	NA
143	715	NA	NA	NA	NA	NA	NA
144	35	50	1.43	44	6.5	0.19	100
145	82	NA	NA	NA	NA	NA	NA
146	37	38	1.03	89	4.8	0.13	131
147	76	158	2.08	13	54	0.71	4
148	38	26	0.68	133	8.2	0.22	88
149	53	60	1.13	78	11	0.21	94
150	31	NA	NA	NA	NA	NA	NA
151	66	NA	NA	NA	NA	NA	NA
152	40	28	0.70	131	3.3	0.08	159
153	43	37	0.86	110	5.3	0.12	135
154	67	65	0.97	96	18	0.27	64
155	127	126	0.99	93	27	0.21	89
156	99	125	1.26	65	36	0.36	40
157	67	133	1.99	15	44	0.66	7
158	65	42	0.65	139	6.1	0.09	148
159	48	NA	NA	NA	NA	NA	NA
160	43	31	0.72	128	5.6	0.13	129
161	64	36	0.56	148	7.1	0.11	138
162	95	84	0.88	106	16	0.17	109
163	68	29	0.43	165	4.2	0.06	165
164	83	115	1.39	48	17	0.20	95
165	44	26	0.59	144	3.6	0.08	160
166	86	159	1.85	23	39	0.45	23
167	47	20	0.43	166	2.8	0.06	166
168	85	60	0.71	129	8.1	0.10	147
169	23	NA	NA	NA	NA	NA	NA
182	48	10	0.21	178	1.36	0.03	176
183	28	27	0.96	98	5.9	0.21	92
184	37	58	1.57	39	16	0.43	26

Institutional and Per Capita Grant Support *(Continued)*

Code	FASEB Members	No. of Grants	G/F	Rank ($N = 181$)	Funding M$	M$/F	Rank ($N = 181$)
185	58	34	0.59	145	6.3	0.11	140
190	28	13	0.46	159	2.7	0.10	145
192	41	NA	NA	NA	NA	NA	NA
193	73	77	1.05	88	17	0.23	81
194	37	27	0.73	127	4.3	0.12	136
195	10	16	1.60	35	2.3	0.23	82
198	22	21	0.95	100	3.7	0.17	110
199	40	NA	NA	NA	NA	NA	NA
206	36	69	1.92	21	14	0.39	34
208	111	83	0.75	125	19	0.17	107
209	24	12	0.50	157	4.1	0.17	108
211	23	17	0.74	126	3.7	0.16	115
212	32	49	1.53	40	9.4	0.29	56
213	32	12	0.38	171	8.2	0.26	69
214	69	53	0.77	123	11	0.16	116
217	31	36	1.16	75	7.7	0.25	75
218	36	48	1.33	54	8.8	0.24	77
220	27	71	2.63	4	18	0.67	6
221	34	NA	NA	NA	NA	NA	NA
222	18	23	1.28	64	3.3	0.18	103
223	21	14	0.67	135	2.3	0.11	139
224	19	108	5.68	1	50	2.63	1
Totals	19,934	24,270	1.04		5897.88	0.24	

Appendix V

Information Sources and Database Access

The NIH Internet Gopher server is a network-based computer service that distributes information about NIH health and clinical issues, NIH-funded grants and research projects, and a variety of research resources. This source can provide the most up-to-date information on the research funding of specific investigators and institutions as well as an on-line listing and descriptions of currently available academic positions through the Academic Position Network (APN) and Academic Physician and Scientist. The provision of interactive log-in to a remote host is a basic Internet service. Telnet is both a protocol and a program that enables you to do so. Instructions for users who don't have an NIH Convex account, but who have Internet connectivity and can run their own Gopher "client" program:

1. Log-in to your own local internet account. The exact procedure varies for each user.
2. Tunnel to the NIH Gopher by typing the command GOPHER. NIH.GOV 70.
3. From menu, select the desired item.

Instructions for users who have Internet connectivity but do not have their own Gopher client. (Internet users who have Telnet capabilities can connect to a "public" gopher site and from this site they can "tunnel" over to the NIH server.)

1. Telnet to a public Gopher site, e.g., consultant.micro.umn.edu. In this example, log-in as gopher.
2. From menu, select: Other Gopher and Information Servers/
3. From menu, select: North America/
4. From menu, select: USA/
5. From menu, select: general/ (not Maryland)

6. From menu, select: National Institutes of Health/
7. From menu, select: desired item
8. Information in the text files and the "About" files provide instructions for using this on-line information.

PUBLIC ACCESS GOPHER SITES

Following are public log-ins for Gopher. It is recommended that you run the client software instead of logging into the public log-in sites. A client uses the custom features of the local machine (mouse, scroll bars, etc.) and gives faster response. The following lists provide the host name, the IP#, and the log-in.

United States

consultant.micro.umn.edu 134.84.132.4 gopher
seymour.md.gov 128.8.10.46 gopher
gopher.msu.edu 35.8.2.61 gopher
twosocks.ces.ncsu.edu 152.1.45.21 gopher
cat.ohiolink.edu 130.108.120.25 gopher
ENVIROLINK.hss.cmu.edu (password; envirolink)
wsuaix.csc.wsu.edu 134.121.1.40 wsuinfo
telnet.wiscinfo.wisc.edu wiscinfo
scilibx.ucsc.edu 128.114.143.4 INFOSLUG
infopath.ucsd.edu INFOPATH
sunsite.unc.edu 152.2.22.81 gopher
uxl.cso.uiuc.edu 128.174.5.59 gopher
panda.uiowa.edu 128.255.40.201 gopher
inform.umd.edu 128.8.10.29 gopher
grits.valdosta.peachnet.edu 131.144.8.206 gopher
gopher.virginia.edu 128.143.22.36 gwis
ecosys.drdr.virginia.edu 128.143.86.233 gopher
gopher.ORA.com 140.186.65.25 gopher
gopher.internet.com 192.215.1.51 enews

Other Countries

nicol.jvnc.net 128.121.50.2 NICOL
finfo.tu-graz.ac.at 129.27.2.4 info

info.anu.edu.au 150.203.84.20 info
nstn.ns.ca 137.186.128.11 fred
camsrv.camosun.bc.ca 134.87.16.4
tolten.puc.cl 146.155.1.16 (Chile)
gopher.denet.dk 129.142.6.66 (Denmark)
gopher.th-darmstadt.de 130.83.55.75
ecnet.ec 157.100.45.2 (Ecuador)
gopher.uv.es 147.156.1.12 (Spain)
gopher.isnet.is 130.208.165.63 (Iceland)
saim.mi.cnr.it 155.253.1.40 (Italy)
gopher.torun.edu.pl 158.75.2.5 (Poland)
sunic.sunet.se 192.36.125.2 (Sweden)
gopher.chalmers.se 129.16.221.40 (Sweden)
hugin.ub2.lu.se 130.235.162.12 (Sweden)
info.brad.ac.uk 143.53.2.5 info

Notes

Introduction

[1] E. B. Fiske, "Major faculty shortages forseen for the 1990s," *The New York Times*, Sept. 13, 1989, 11.

[2] The response rates for the survey were 32.4% for U.S. medical school faculty and 33.4% for staff at other institutions: there were no fewer than 3 respondents per school. This response compares very favorably with the 20% rate usually achieved in mail surveys. The respondents from U.S. medical schools were 90% white, 83% male, 70% tenured, and 52% full professors. The corresponding demographics for other faculty and staff were 91% white, 81% male, 70% tenured, and 49% full professors. These characteristics closely correspond to the profile for U.S. scientists and engineers (Survey of Doctorate Recipients, National Research Council, 1987). Most (56% at U.S. medical schools and 59% at other institutions) had been employed at their current institution for more than 10 years. They worked an average of 40–60 hour/week (70% and 72%), devoting more than one-half of their total time to research (60% and 62%) and less than one-fifth to teaching (56% and 51%). Most (57% and 50%) had no clinical or other responsibilities, e.g., consulting.

[3] *The Best of Confucius*, tr. by J. R. Ware (Garden City, NJ: Halcyon House, 1950).

Chapter 1. The Offer

[1] Carl J. Sinderman has also analyzed the personality types of science administrators in *Winning the Games Scientists Play* (New York: Plenum, 1982), 223 et seq. and *Survival Strategies for New Scientists* (New York: Plenum, 1987), 183 et seq.

Chapter 2. The Contract

[1] American Chemical Society Annual Salary Survey (Washington, DC: American Chemical Society, 1988).

Chapter 3. Lawsuits

[1] Will *v*. Michigan Dep't. of State Police, 109 S.Ct. 2304.

[2]Equal Employment Opportunity Commission, *Guidelines Defining Sexual Harassment under Title VII of the Civil Rights Act of 1964* (Washington, DC: Equal Employment Opportunity Commission, 1980).

Chapter 5. Training

[1]H. A. Krebs, "The making of a scientist," *Nature* 215(1967): 1441–1445.

[2]Kanigel, *Apprentice to Genius* (Macmillan: New York, 1986).

[3]J. R. McIntosh, "Funding constraints and population growth: The cell biologist's nightmare," *ASCB Newsletter* 17(1994):1; "The population dilemma," *ASCB Newsletter* 18(1995):1; T. J. Kennedy, "Graduate education in the biomedical sciences: Critical observations on training for research careers," *Academic Medicine* 69(1994): 779–799; J. M. Musacchio, "American science in crisis: The need to revise the NIH funding policy," *FASEB Journal* 8(1994): 679–683.

[4]D. G. Brown, *The Mobile Professors* (Washington D.C.: American Council on Education, 1967), 140.

Chapter 6. U.S. Medical Schools

[1]Lanks, K. W. *Institutional Subjective Factors Survey*. (New York: Faculty Press, 1989).

[2]*SAS User's Guide, Statistics Version*, 5th Ed. (Gary, N.C.: SAS Institute, Inc., 1985)

Chapter 7. Alternatives

[1]S. C. Carrier, and D. Davis-Van Atta, "Educating America's Scientists: The Role of Research Colleges" in "Doing Science Off the Beaten Path at Liberal Arts Schools," *The Scientist*, November 23(1992):21–23.

Chapter 8. Productivity

[1]A. E. Bayer and J. E. Dutton, "Career age and research-professional activities of academic scientists: Tests of alternative non-linear models and some implications for higher education faculty policies," *Journal of Higher Education* 48(3)(1977): 259–282. Recent articles and number of citations in Science Citation Index were among the eight measures of productivity used in this study. Productivity of an academic institution can be defined as the ratio of input to output ("Measuring and Increasing Academic Productivity" in: *New Directions for Institutional Research*, Vol. II, No. 4, Issue Ed. R. A. Wallhaus). In these terms, resources need to maintain each FASEB faculty member is the input. Of course, grants and publications are only two of many possible measures of output, e.g., credit hours taught or doctorates granted.

[2]J. Gourman, *The Gourman Report*, 4th Ed. (Los Angeles: National Education Standards, 1987).

[3]J. L. Bowitch and A. F. Buono, *Quality of Work Life Assessment: A Survey-Based Approach* (Boston: Auburn House, 1982).

[4]R. Likert, *Motivation: The Core of Management* (New York: American Management Associations, Personnel Series No. 155, 1953), 3–20.

[5]*World Directory of Medical Schools* (Geneva: World Health Organization, 1988); *American Universities and Colleges*, 13th Ed. (New York: Walter DeGruyter, 1987).

[6]*Directory of Members 1993/1994.* (Bethesda, MD: Federation of American Societies for Experimental Biology, 1993).

[7]*National Institutes of Health Research Grants. Fiscal Year 1993* (NIH GOPHER Database).

[8]*SAS User's Guide, Statistics Version*, 5th Ed. (Gary, N.C.: SAS Institute, Inc., 1985).

[9]M. S. N. Rao and G. Kegeles, "An ultracentrifuge study of the polymerization of a-chymotrypsin," *Journal of the American Chemical Society* 80(1958): 5724–5729.

Chapter 9. Productivity Versus Atmosphere

[1]This phase of the study was performed with the help of Ping-Wu Li of the Scientific Computing Center, State University of New York Health Science Center at Brooklyn.

[2]Survey responses to questions 4, 6, 25, 29, and 39 were highly correlated with production of grants and publications, but were represented most strongly in the academic atmosphere factor group. Priority was given to factor group representation.

[3]D. G. Brown, *The Mobile Professors* (Washington, D.C.: American Council on Education, 1967), 162.

[4]A similar conclusion, that developing a corporate culture conducive to productivity is a key management function, is also proposed by T. J. Peters and R. H. Waterman in *In Search of Excellence* (New York: Harper & Row, 1982).

Chapter 10. Quality

[1]A. E. Bayer and J. E. Dutton, "Career age and research-professional activities of academic scientists: tests of alternative non-linear models and some implications for higher education faculty policies," *Journal of Higher Education* 48(3)(1977): 259–282.

[2]These data were obtained from the collected works of Eugene Garfield, *Essays of an Information Scientist, Vols. 1–8* (Philadelphia: ISI Press, 1977–1985).

[3]The regression analysis was performed in two ways: using the raw number of highly cited publications and using a surrogate variable assigned after grouping the number of highly cited publications into categories (0 = 1), (1,2 = 1), (3 through 9 = 3), (≥ 10 = 4). Survey responses found to be correlated by either method are listed together.

Index

T - #0594 - 101024 - C0 - 229/152/8 - PB - 9781560324232 - Gloss Lamination